FORMULAS AND TABLES

FOR

ARCHITECTS AND ENGINEERS

IN

CALCULATING THE STRAINS AND CAPACITY

OF

STRUCTURES IN IRON AND WOOD,

BY

F. SCHUMANN, C. E.

ILLUSTRATED WITH MORE THAN THREE HUNDRED DIAGRAMS, DESIGNED AND ENGRAVED ESPECIALLY FOR THIS WORK BY J. C. LYONS.

WASHINGTON CITY:
WARREN CHOATE & CO.
1873.

STEREOTYPED BY
M'GILL & WITHEROW,
WASHINGTON, D. C.

THIS VOLUME

IS

RESPECTFULLY DEDICATED

TO

A. B. MULLETT,

SUPERVISING ARCHITECT OF THE U. S. TREASURY DEPARTMENT,

BY THE AUTHOR.

PREFACE.

The following collection of Formulas and Tables is intended as a ready reference for Architects and Engineers in computing the Strains and Strength of Structures in Iron or Wood. The author has attempted to give concisely all the necessary data required for the calculations. The object is, to give a compilation and transformation into simple formulas and tables, *adapted to practice*, of matter contained in the works of *Weisbach*, *Rankine*, *Rebhahn*, *Ritter*, *Breyman*, *Gordon*, *Brandt*, *Moll & Reuleaux*, *Laissle & Schübler*, *Fairbairn*, and others.

Among other matter may be particularly mentioned the tables giving the capacity of rolled and cast-iron beams, and the extensive table of formulas for the Moment of Inertia and Resistance of various sections. The formulas and constants for ascertaining the Strains in Roof and other Trusses will be found quite complete, extending to all the different forms or systems in general use. Under "Miscellaneous" is given necessary formulas for the calculation of Lines, Areas, and Solids; also tables of Natural Sines, Cosines, Tangents, Cotangents, Secants, and Cosecants, Area and Circumference of Circles, Weight of Materials, &c.

WASHINGTON, D. C., *March*, 1873.

ERRATA.

On page 4, 10th line from bottom, read $\frac{30}{100}$ instead of 30.

On page 4, 10th line from bottom, read 10.0036 instead of 10.036.

On page 4, 14th, 15th, and 16th lines from bottom, read $\frac{a}{100}$ instead of a.

On page 32, *Fig.* 70, insert $l =$ distance between supports.
On page 34, *Fig.* 72, insert $l =$ distance between supports.
On page 34, *Fig.* 74, insert $l =$ length of beam.
On pages 38 and 39 $w =$ total weight of beam between supports.
On page 39, 5th line from top, read 1099000 instead of 1000000.
On page 39, 5th line from top, read 1754 instead of 1757.
On pages 144 and 145, in formulas for H_n, change places of last minus sign with foregoing plus sign. (See 13th line from top.)

Page 145, lines 1 to 7 from bottom, Page 146, lines 1 to 3 from top, Page 146, lines 13 to 22 from top,	Change places of C and T under strains in *Figs.* 225, 226, 227, and 228.

On page 149, 1st line from bottom, read $\frac{lw}{N} \cdot \frac{D}{H-D}$ instead of $\frac{lw}{N}$.

On page 197, 7th line from bottom, read 3.14159 instead of 1.14159.
On page 204, 1st line from bottom, read $A + A_{\prime}$ instead of $AA_{\prime}$.

CONTENTS.

MISCELLANEOUS.

FORMULAS AND TABLES

FOR

ARCHITECTS AND ENGINEERS.

Summary of Definitions and General Principles.

EXTERNAL FORCES are those forces (loads, &c.) which cause or tend to cause the rupture of a structure.

INTERNAL FORCES are those forces which resist the external forces; when one balances the other, the structure is said to possess *Stability*.

EXTERNAL FORCES.	INTERNAL FORCES.
Compressive strain.	Resistance to Compression.
Tensional strain.	Resistance to Tension.
Shearing strain.	Resistance to Shearing.
Cross-breaking strain.	Resistance to Cross-breaking.

COMPRESSION causes the material to fail by *crushing* or *buckling*, or both.

RESISTANCE *to direct Crushing:* In case pillars, blocks, struts, or rods, along which the strains act, are not so long in proportion to their diameter as to have a tendency to give way by bending sideways. This includes—

Stone and brick pillars and blocks, of ordinary proportions;

Pillars, struts, and rods of cast iron, in which the length is not more than five times the diameter, approximately;

Pillars, struts, and rods of wrought iron, in which the length is not more than ten times the diameter, approximately;

Pillars, struts, and rods of dry timber, in which the length is not more than twenty times the diameter.

Let $W =$ Crushing load in lbs.
$C =$ Ultimate resistance of material to crushing in lbs. per square inch.
$A =$ Sectional area of pillar, &c., in square inches.

Then will $W = A \times C$; and $A = \dfrac{W}{C}$

TENSION, causes the material to be torn asunder.

Resistance of bars, &c., to teaiing: the ultimate strength of a bar (to tearing) is: when

$T =$ Ultimate resistance of the material to tearing, in lbs. per square inch.
$W =$ Tearing load in lbs.
$A =$ Sectional area of bar, in square inches.

Then will $W = A \times T$; and $A = \frac{W}{T}$

SHEARING causes the fibres of the material to be shorn by each other; when a bolt pulls out of its nut, the threads of the screw are said to be sheared.

Resistance of bars, bolts, &c., when sheared at one place, is: when

$S =$ Ultimate resistance of material to shearing, in lbs. per square inch.
$W =$ Shearing load in lbs.
$A =$ Sectional area of bar, &c., in square inches.

Then will $W = A \times S$; and $A = \frac{W}{S}$

CROSS-BREAKING a beam, &c., supported at one or both ends, with a force at right angles to its length, sufficient to cause rupture, is said to be broken across.

Resistance to cross-breaking is the resistance of the material to compression, tension, and shearing combined; ———.

The flanges or booms, in beams or trusses, resist the bending moment, or moment of rupture.

The web or braces, in beams or trusses, resist the shearing forces.

CAPACITY means the load or pressure a structure is capable of sustaining with safety.

DEFLECTION is the displacement of a beam from a horizontal, caused by its own weight or the applied load, or both.

CAMBER is given a beam to counter balance the deflection, so that the beam may be horizontal when loaded; the camber should be equal to the computed deflection.

To find the effect of combining several loads on one beam, whose separate actions are known: for each cross section, the shearing force is the sum of the shearing forces, and the bending moment the sum of the bending moments, which the loads would produce separately.

When a load on a structure is partly static and partly moving, multiply each part of the load by its proper factor of safety, and add together the products: the sum will be the load to which the structure is to be adapted.

For a Bridge with two platforms, one carrying a road and the other a railway, those two loads are to be combined.

Of Iron Ties, Struts, and Beams.

In designing ordinary structures of wrought iron, it saves time and expense to use iron bars of such forms of cross section as are usually to be met with in the market. An engineer should avoid introducing new sections for bars into his designs, except when, by so doing, some important purpose is to be served, or some decided advantage to be gained. The most common forms of rolled bars is shown by the following enumerated figures:

No. of figure.	Name of Form.	Applicable for—
4	Square iron	Ties.
13	Round iron	Ties, bolts, and rivets.
2	Flat iron	Ties.
29	I or double T-iron	Beams, rafters, and struts.
30	Channel iron	Rafters and struts.
37	T-iron	Rafters and struts.
47	L or angle iron	Various purposes.
1	Deck Beam	Beams and rafters.

Avoid the use of cast iron for ties, also trussed cast-iron beams.

When a member of a structure acts alternately as a strut and as a tie, it must have sufficient total sectional area, and sufficient stiffness, to resist the greatest compressive strain that can act, and sufficient effective sectional area to resist the greatest tensional strain which can act along it.

Let all pins, bolts, rivets, &c., exposed to a shearing strain, fit tight in its hole or socket.

Roof trusses, the divisions of a rafter, and also the struts, may be considered as hinged at the ends.

In members under a compound strain, as for instance a trussed beam with a distributed load, be careful to take into account the additional compression, caused by the bending moment.

The best distribution of the material in a section of a cast-iron beam, for cross-breaking, is that $\frac{T}{s} = \frac{C}{s_{\prime}}$; or $\frac{s_{\prime}}{s} = \frac{C}{T}$

When $T =$ Ultimate resistance of the material to tension.
$C =$ Ultimate resistance of the material to compression.
$s =$ Distance from neutral axis to most extended fibres.
$s_{\prime} =$ Distance from neutral axis to most compressed fibres.

That is, the fibres most in tension should be nearest the neutral axis of beam.

In wrought-iron beams, the section may be made alike above and below the neutral axis.

THE MODULUS OF RUPTURE should be applied to beams with full section, or beams with continuous web only; for all open web beams, or beams with very thin web, the resistance of the material to crushing or tearing, respectively, must be used instead.

EXPANSION AND CONTRACTION of long beams, which arise from the changes of atmospheric temperature, are usually provided for by supporting one end of each beam on rollers of steel or hardened cast iron. The following table shows the proportions in which the length of a bar of certain materials is increased by an elevation of temperature from the melting point of ice (32° Fahr., or 0° Centigrade) to the boiling point of water under the mean atmospheric pressure, (212° Fahr., or 100° Cent.;) that is, by an elevation of 180° Fahr., or 100° Cent.:

METALS.		EARTHY MATERIALS.	
Brass	0.00216	Brick, common	0.00355
Bronze	0.00181	Brick, fire	0.00050
Copper	0.00184	Cement	0.00140
Cast iron	0.00111	Glass, average	0.00090
Wrought iron	0.00120	Granite	0.00085
Tin	0.00225	Marble	0.00087
Zinc	0.00294	Sandstone	0.00105
Lead	0.00290	Slate	0.00104

Reference.

Let $u =$ Value given in above table, for a certain material.
$l =$ Length of a bar at 0° Centigrade,
and $l_{,} =$ its length at a given number of degrees Centigrade.
$a =$ Given number of degrees, at which $l_{,}$ is required.
$A =$ Superficial area of a plate;
and $A_{,} =$ its area at a given number of 0° C.
$B =$ Cubic contents of a body,
and $B_{,} =$ its contents at a given number of 0° C.

Then will $l_{,} = l\ (1 + a\,u)$;
$A_{,} = A\,(1 + 2\,a\,u)$;
$B_{,} = B\,(1 + 3\,a\,u)$.

Example: A bar of wrought iron 2 inches square, is 10 feet long at a temperature of 0° Centigrade; what is its length at an increased temperature of 30°?

Ans: $l_{,} = 10\,(1 + 30 \times 0.00120) = 10.036$ feet.

THE NEUTRAL AXIS, in a cross section of a beam, is that layer of fibres which are neither in compression or tension, by the action of a load on the beam; it always passes through the centre of gravity of the section: provided the limits of elasticity of the material is not exceeded. A beam supported at both ends, with a load acting perpendicular between the supports, will cause the fibres above the neutral axis to be compressed, and those below, extended: the farther from the fibres to the neutral axis, the greater the stress.

Under MOMENT OF INERTIA of a cross section, may be understood: an internal force at rest; a static force resisting an external force; it means the sum of all the area elements, multiplied by the square of their perpendicular heights from the neutral axis of the section. The moment of inertia, which we have denoted with I, depends on the form and dimensions of the cross section, and is expressed in square inches.

MOMENT OF RESISTANCE of a cross section is that static force resisting an external force of compression or tension; it is equal to the moment of Inertia divided by the distance of the most extended or compressed fibres, respectively, from the neutral axis.

MOMENTS OF INERTIA AND RESISTANCE OF VARIOUS SECTIONS.

To find the moment of inertia of any given cross section—

FIRST. Divide the section into as many simple figures as possible. (See diagram, fig. 1.)

SECOND. Find the moment of inertia of each of the simple figures about its own neutral axis, and insert the value in the following formula:

Reference.

Letters $A, A_{/}, A_{//}$, = area of simple figure, respectively; and
$d, d_{/}, d_{//}$, = its distance from its centre of gravity to that of the whole section.
$i, i_{/}, i_{//}$, = moment of inertia of simple figures, respectively.

For neutral axis see centre of gravity.

Fig. 1.

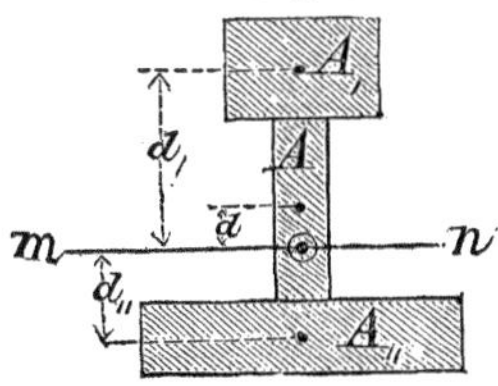

Formula.

$I = (i + d^2A) + (i_{/} + d_{/}^2A_{/}) + (i_{//} + d_{//}^2A_{//}) +$ &c., = moment of inertia of whole section.

MOMENTS OF INERTIA I AND MOMENTS OF RESISTANCE $\frac{I}{s}$

Reference.

$m - n$ = neutral axis of section.
r = radius.
s = distance from neutral axis to most compressed or extended fibres.
b, h, &c. = dimensions.
A = area.

No. of Section.	No. of Figure.	Form of Section.
I.	2 and 3	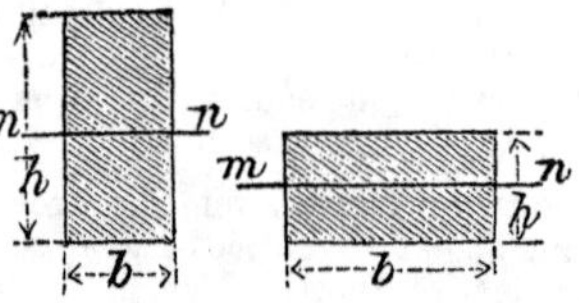 m n h b — m n h b
II.	4	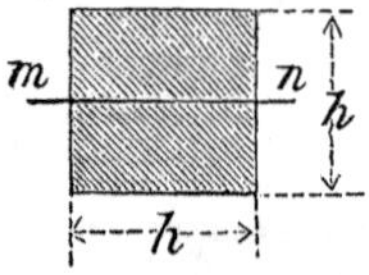 m n h h
III.	5	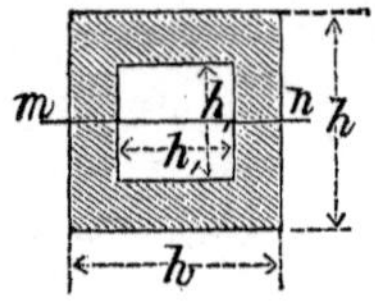 m h_1 n h h_1 h
IV.	6	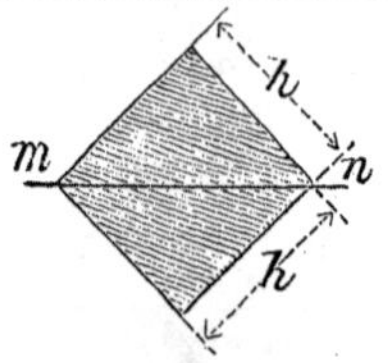 h m n h
V.	7	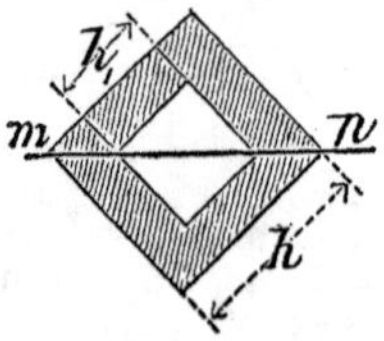 h_1 m n h

Moment of Inertia I.	Moment of Resistance $\frac{I}{s}$
$\frac{1}{12} bh^3 = \frac{1}{12} Ah^2$	$\frac{bh^2}{6}$
$\frac{1}{12} h^4 = \frac{1}{12} Ah^2$	$\frac{h^3}{6}$
$\frac{h^4 - h_{\prime}^4}{12}$	$\frac{h^4 - h_{\prime}^4}{6h}$
$\frac{1}{12} h^4 = \frac{1}{12} Ah^2$	$0.118h^3$
$\frac{h^4 - h_{\prime}^4}{12}$	$\frac{h^4 - h_{\prime}^4}{12h} \cdot \sqrt{2}$

No. of Section.	No. of Figure.	Form of Section.
VI.	8	b, m, n, h
VII.	9	$b_{,}$, m, n, h, $h_{,}$, b
VIII.	10	m, n, h, b
IX.	11	h, g
X.	12	m, n, h, b

Moment of Inertia I.	Moment of Resistance $\frac{I}{s}$
$\frac{1}{48} bh^3$	$\frac{1}{24} bh^2$
$\frac{1}{48} (bh^3 - b_{/}h_{/}^3)$	$\frac{1}{24} \frac{bh^3 - b_{/}h_{/}^3}{h}$
$\frac{1}{48} bh^3 = \frac{1}{24} Ah^2$	$\frac{1}{24} bh^2$
$\frac{1}{36} gh^3 = \frac{1}{18} Ah^2$	$\frac{1}{24} gh^2$
$\frac{1}{48} bh^3 = \frac{1}{24} Ah^2$	$\frac{1}{24} bh^2$

No. of Section.	No. of Figure.	Form of Section.
XI.	13	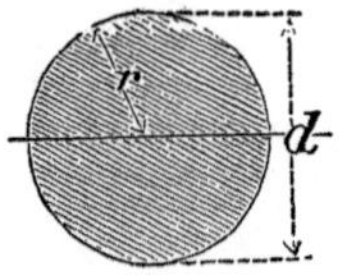
XII.	14	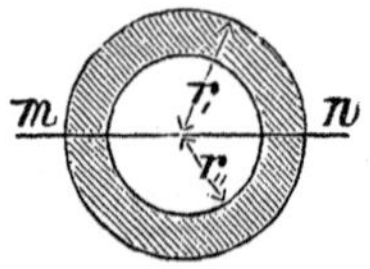
XIII.	15	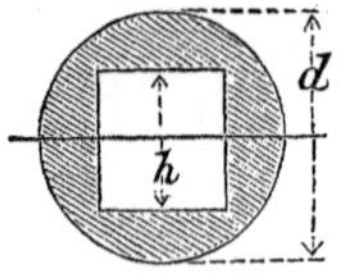
XIV.	16	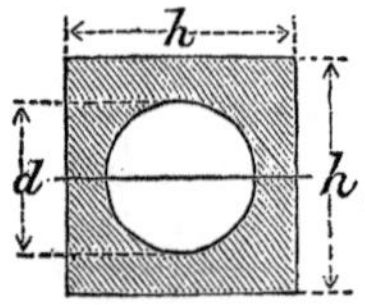
XV.	17 and 18	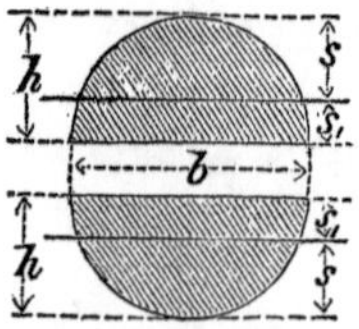

Moment of Inertia I.	Moment of Resistance $\frac{I}{s}$
$\frac{1}{4}\pi r^4 = \frac{1}{16} Ad^2$	$\frac{1}{4}\pi r^3 = \frac{1}{4} Ar$
$\frac{1}{4}\pi (r_{\prime}^4 - r_{\prime\prime}^4)$	$\frac{1}{4}\pi \frac{r_{\prime}^4 - r_{\prime\prime}^4}{r_{\prime}}$
$\frac{\pi}{64} d^4 - \frac{h^4}{12} = 0.0491 d^4 - \frac{h^4}{12}$	$\frac{I}{\frac{1}{2} d}$
$\frac{h^4}{12} - \frac{\pi}{64} d^4 = \frac{h^4}{12} - 0.0491 d^4$	$\frac{I}{\frac{1}{2} h}$
$\frac{12}{175} \cdot Ah^2 = \frac{8}{175} bh^3$	$s = 0.576h = (1 - \frac{4}{3\pi}) h$ $s_{\prime} = 0.424h = \frac{4}{3\pi} h$

No. of Section.	No. of Figure.	Form of Section.
XVI.	19	
XVII.	20	
XVIII.	21	
XIX.	22	
XX.	23	

Moment of Inertia I.	Moment of Resistance $\frac{I}{s}$
$\frac{1}{30}\,bh^3 = \frac{1}{20}\,Ah^2$	$\frac{bh^2}{15} = \frac{1}{10}\,Ah$
$\frac{1}{64}\,\pi\,bh^3 = \frac{1}{16}\,Ah^2$	$\frac{1}{32}\,\pi\,bh^2 = \frac{1}{8}\,Ah$
$\frac{8}{175}\,bh^3 = \frac{12}{175}\,Ah^2$	$\frac{8}{105}\,bh^2 = \frac{4}{35}\,Ah$
$\frac{1}{30}\,bh^3 = \frac{1}{20}\,Ah^2$	$\frac{bh^2}{15} = \frac{1}{10}\,Ah$
$\frac{\pi}{64}\,bh^3 = \frac{1}{16}\,Ah^2$	$\frac{1}{32}\,\pi\,bh^2 = \frac{1}{8}\,Ah$

No. of Section.	No. of Figure.	Form of Section.
XXI.	24	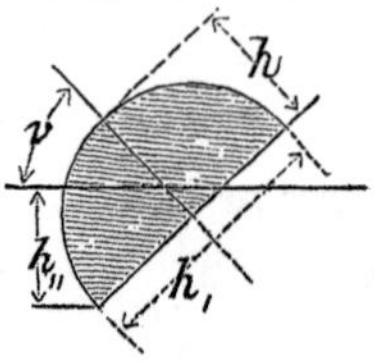
XXII.	25	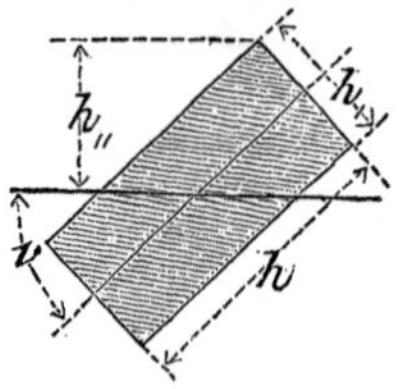
XXIII.	26	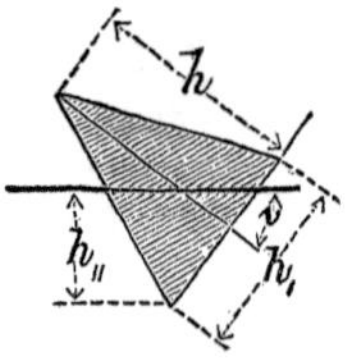
XXIV.	27	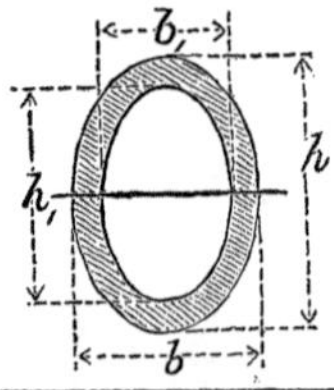
XXV.	28, 29, and 30	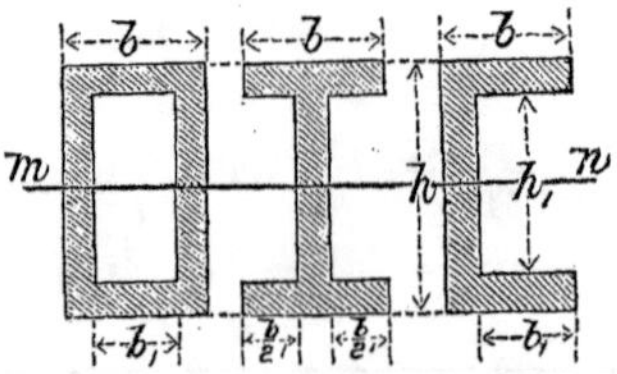

Moment of Inertia I.	Moment of Resistance $\frac{I}{s}$
$\frac{1}{5} A \left[\frac{1}{4} h_{,}^{2} \cos^2 v + \frac{12}{35} h^2 \sin^2 v\right]$	$\frac{I}{h_{,,}}$
$\frac{1}{12} A \left[h^2 \cos^2 v + h_{,}^{2} \sin^2 v\right]$	$\frac{I}{h_{,,}}$
$\frac{1}{6} A \left[\frac{1}{4} h_{,}^{2} \cos^2 v + \frac{1}{3} h^2 \sin^2 v\right]$	$\frac{I}{h_{,,}}$
$\frac{1}{64} \pi (bh^3 - b_{,}h_{,}^{3})$	$\frac{I}{\frac{1}{2} h}$
$\frac{bh^3 - b_{,}h_{,}^{3}}{12}$	$\frac{bh^3 - b_{,}h_{,}^{3}}{6h}$

No. of Section.	No. of Figure.	Form of Section.
XXVI.	31	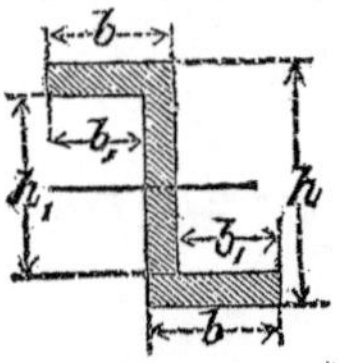
XXVII.	32	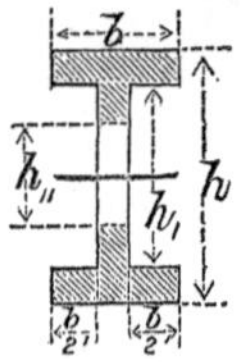
XVIII.	33	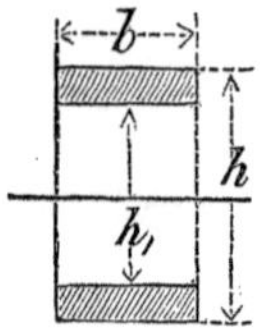
XXIX.	34	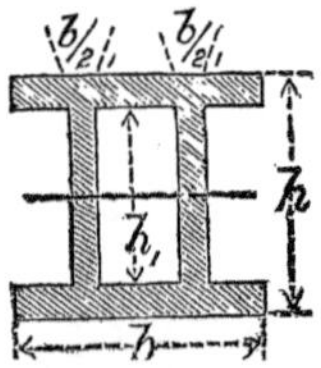
XXX.	35	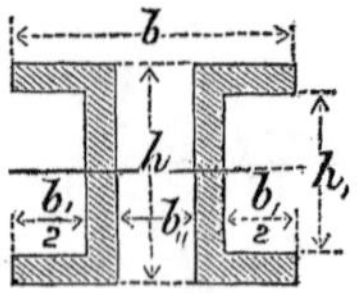

Moment of Inertia I.	Moment of Resistance $\frac{I}{s}$
$\frac{bh^3 - b_{,}h_{,}^3}{12}$	$\frac{bh^3 - b_{,}h_{,}^3}{6h}$
$\frac{1}{12}[bh^3 - b_{,}h_{,}^3 - (b-b_{,})\,h_{,,}^3]$	$\frac{1}{6h}[bh^3 - b_{,}h_{,}^3 - (b-b_{,})\,h_{,,}^3]$
$\frac{1}{12}\,b\,[h^3 - h_{,}^3]$	$\frac{b\,(h^3 - h_{,}^3)}{6h}$
$\frac{1}{12}[bh^3 - bh_{,}^3 + b_{,}h_{,}^3]$	$\frac{1}{6h}\,[bh^3 - bh_{,}^3 + b_{,}h_{,}^3]$
$\frac{1}{12}[(bh^3 - b_{,}h_{,}^3) - (b_{,,}h^3)]$	$\frac{(bh^3 - b_{,}h_{,}^3) - (b_{,,}h^3)}{6h}$

No. of Section.	No. of Figure.	Form of Section.
XXXI.	36 and 37	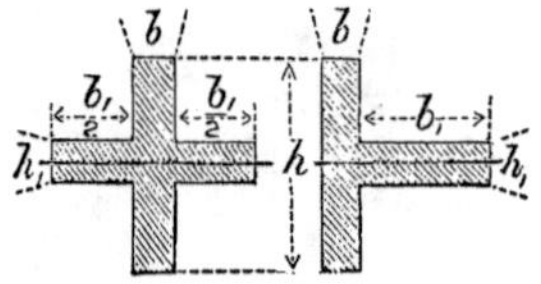
XXXII.	38	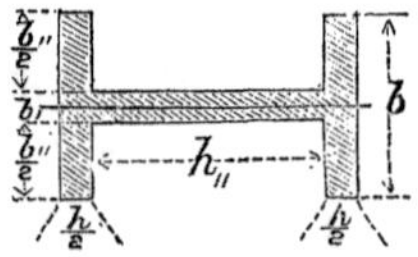
XXXIII.	39	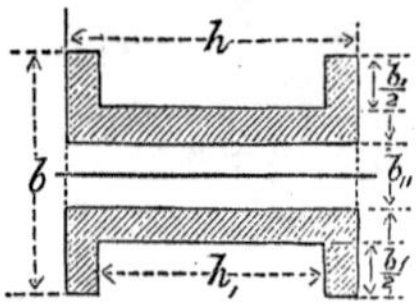
XXXIV.	40	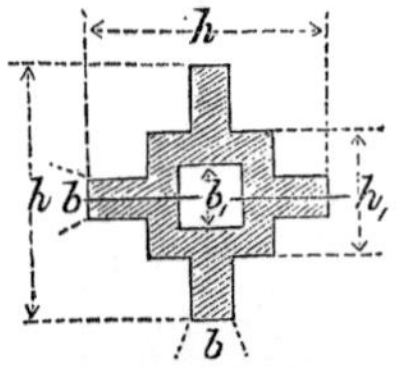
XXXV.	41	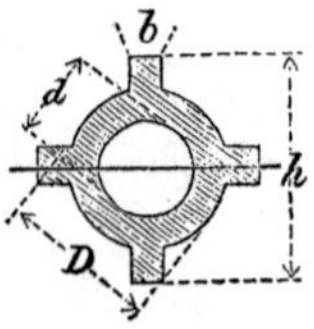

Moment of Inertia I.	Moment of Resistance $\frac{I}{s}$
$\frac{1}{12}(bh^3 + b_{,}h_{,}^3)$	$\frac{bh^3 + b_{,}h_{,}^3}{6h}$
$\frac{1}{12}(hb^3 + h_{,,}b_{,}^3)$	$\frac{hb^3 + h_{,,}b_{,}^3}{6b}$
$\frac{1}{12}[(b^3h - 3b^2b_{,}h_{,} + 3bb_{,}^2h_{,} - b_{,}^3h_{,}) - (hb_{,,}^3)]$	$\frac{I}{\frac{1}{2}b}$
$\frac{1}{12}[h_{,}^4 + b(h^3 - h_{,}^3) + (h - h_{,})b^3 - b_{,}^4]$	$\frac{I}{\frac{1}{2}h}$
$\frac{1}{12}[\frac{3}{16}\pi D^4 + b(h^3 - D^3) + (h - D)b^3] - 0.0491d^4$	$\frac{I}{\frac{1}{2}h}$

No. of Section.	No. of Figure.	Form of Section.
XXXVI.	42	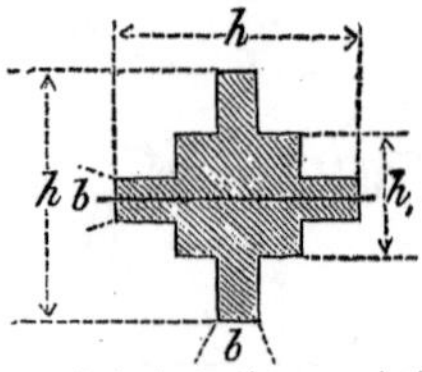
XXXVII.	43	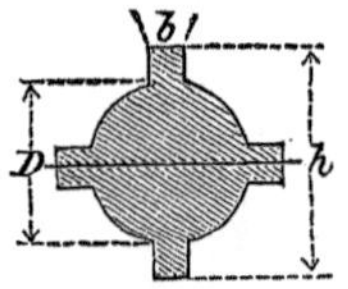
XXXVIII.	44	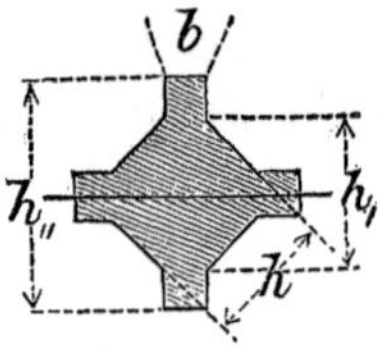
XXXIX.	45	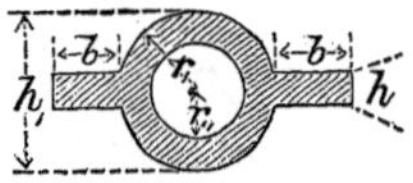
XL.	46, 47, and 48	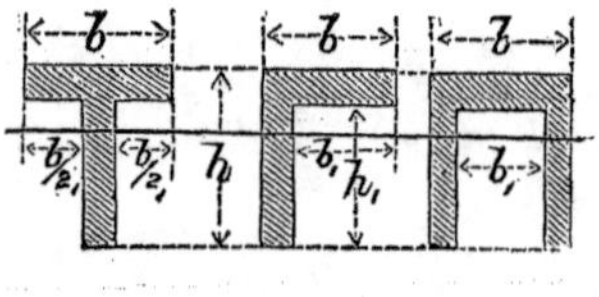

Moment of Inertia I.	Moment of Resistance $\frac{I}{s}$
$\frac{1}{12}[h_{\prime}^4 + b(h^3 - h_{\prime}^3) + (h - h_{\prime})b^3]$	$\frac{I}{\frac{1}{2}h}$
$\frac{1}{12}[\frac{3}{16}\pi D^4 + b(h^3 - D^3) + (h - D)b^3]$	$\frac{I}{\frac{1}{2}h}$
$\frac{1}{12}[h_{\prime}^4 + b(h^3 - h_{\prime}^3) + (h - h_{\prime})b^3]$	$\frac{I}{\frac{1}{2}h_{\prime\prime}}$
$\frac{1}{12}[3\pi(r_{\prime}^4 - r_{\prime\prime}^4) + 2bh^3]$	$\frac{I}{\frac{1}{2}h_{\prime}}$
$\frac{(bh^2 - b_{\prime}h_{\prime}^2)^2 - 4bhb_{\prime}h_{\prime}(h-h_{\prime})^2}{12(bh - b_{\prime}h_{\prime})}$	$\frac{(bh^2 - b_{\prime}h_{\prime}^2)^2 - 4bhb_{\prime}h_{\prime}(h-h_{\prime}h)^2}{6(bh^2 - b_{\prime}h_{\prime}^2)}$

No. of Section.	No. of Figure.	Form of Section.
XLI.	49, 50, and 51	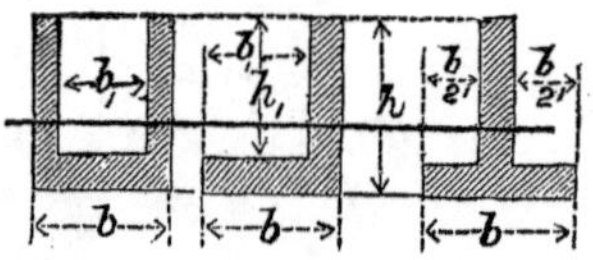
XLII.	52	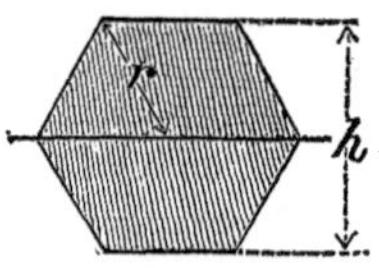
XLIII.	53	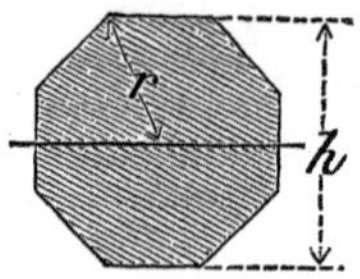
XLIV.	54	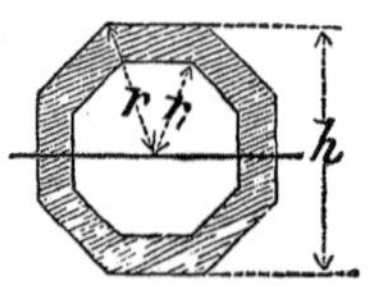
XLV.	55	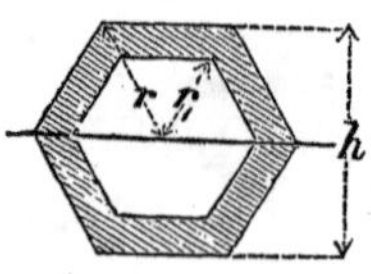

Moment of Inertia I.	Moment of Resistance $\frac{I}{s}$
$\frac{(bh^2 - b_{,}h_{,}^2)^2 - 4bhb_{,}h_{,}(h-h_{,})^2}{12(bh - b_{,}h_{,})}$	$\frac{(bh^2 - b_{,}h_{,}^2)^2 - 4bhb_{,}h_{,}(h-h_{,})^2}{b(bh^2 + b_{,}h_{,}^2 - 2b_{,}hh_{,})}$
$\frac{5}{16} r^4 \sqrt{3} = 0.5413\ r^4$	$\frac{I}{\frac{1}{2}h}$
$\frac{1 + 2\sqrt{2}}{6} r^4 = 0.6381\ r^4$	$\frac{I}{\frac{1}{2}h}$
$\frac{5}{16}\sqrt{3}(r^4 - r_{,}^4)$ $= 0.5413\ (r^4 - r_{,}^4)$	$\frac{I}{\frac{1}{2}h}$
$\frac{1 + 2\sqrt{2}(r^4 - r_{,}^4)}{6}$ $= 0.6381\ (r^4 - r_{,}^4)$	$\frac{I}{\frac{1}{2}h}$

No. of Section.	No. of Figure.	Form of Section.
XLVI.	56	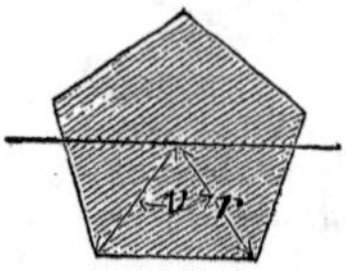
XLVII.	57	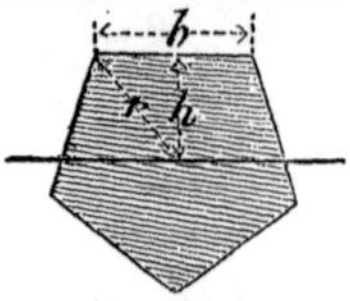
XLVIII.	58	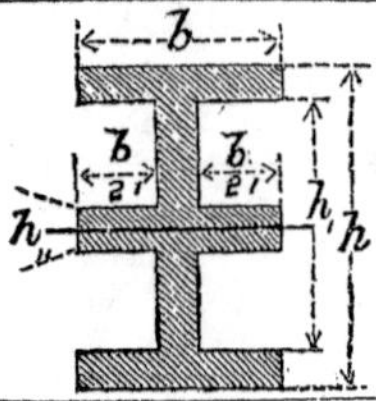
XLIX.	59	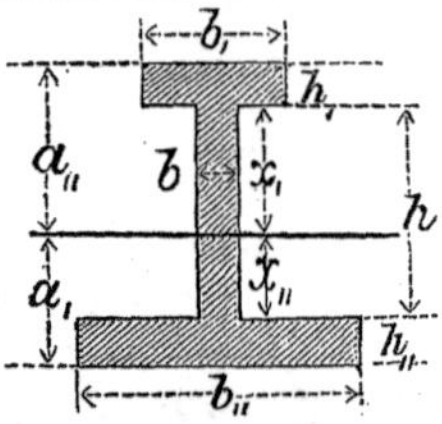
L.	60	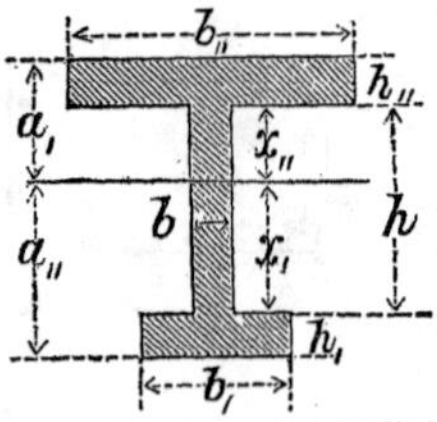

Moment of Inertia I.	Moment of Resistance $\frac{I}{s}$
$n_{,}$ = number of sides. $\frac{1}{24}\, n_{,} r^4\, sin.v\, (2 + cos.v)$	$\frac{1}{24}\, n_{,}\, r^3\, sin.v\, (2 + cos.v)$
$n_{,}$ = number of sides. b = length of a side. $\frac{1}{12}\, A\, (3h^2 + \frac{1}{4}\, b^2)$	$\frac{1}{12}\, \frac{A}{h}\, (3h^2 + \frac{1}{4}\, b^2)$
$\frac{bh^3 - b_{,}h_{,}^3 + b_{,}h_{,,}^3}{12}$	$\frac{bh^3 - b_{,}h_{,}^3 + b_{,}h_{,,}^3}{6h}$
$I = \frac{1}{3} \left\{ \begin{array}{l} b_{,,}\, (a_{,}^3 - x_{,,}^3) + \\ b\ \ (x_{,,}^3 + x_{,}^3) + \\ b_{,}\ \ (a_{,,}^3 - x_{,}^3) \end{array} \right\};$	$\frac{I}{a_{,}}$
$x_{,} = \frac{bh^2 - b_{,}h_{,}^2 + b_{,,}h_{,,}^2 + 2b_{,,}\, hh_{,,}}{2\, (bh + b_{,}h_{,} + b_{,,}\, h_{,,})}$ $x_{,,} = h - x_{,}$ $a_{,} = x_{,,} + h_{,,}$ $a_{,,} = x_{,} + h_{,}$	$\frac{I}{a_{,,}}$

STRENGTH OF MATERIALS, &c.,

In lbs., avoirdupois, per square inch of cross-section.

Materials.	Weight of a cubic foot.	Ultimate Resistance to—				Modulus of elasticity.
		Tearing.	Crushing.	Shearing.	Cross-br'k. Modulus of Rupture.	
METALS.						
Brass, cast, average............	505.7	18000	10300	....	....	9170000
" wire....................	533	49000	....	....	....	14230000
Bronze or gun metal, (copper 8, tin 1)	524	36000	....	....	....	9900000
Copper, cast....................	537	19000	117000			
" sheet....................	549	30000				
" bolts....................	......	35000				
" wire....................	......	60000	....	....	....	17000000
Iron, cast, average..............	445	16500	112000	27700	....	17000000
" various..............	434 to 456	13400 to 29000	80000 to 145000	....	....	14000000 to 22900000
" beams, average...	......	....	....	....	28800	
" " open work.	......	....	....	....	17000	
" solid rect. bars, various quallities.	......	....	....	....	33000 to 43500	
Iron, wrought, average......	481	65000	36000 to 40000	50000		
" beams	......	....	....	....	38000	
" plates........	......	51000				
" joints, d'ble riveted.	......	35700				
Iron, wrought, joints, single riveted.	......	28600				
Iron, wrought, bars and bolts.	......	60000 to 70000	....	....	....	29000000
" hoop, best best	......	64000				
" wire..............	......	70000 to 100000	....	....	....	25300000
" wire ropes....	......	90000	....	....	....	15000000
Steel, average....................	490	....	....	....	80000	
" bars..........................	......	100000 to 130000	12000	15000	....	29000000 to 42000000
" plates..........................	......	80000				
Lead, sheet..........................	712	3300	7730	....	....	720000
Tin, cast..........................	462	4600	15500	....	....	4000000
Zinc..........................	436	7000 to 8000	....	....	....	13000000
TIMBER, (well seasoned and dry.)						
Ash..........................	47	17000	9000	1400	12000 to 14000	1600000
Bamboo..........................	25	6300				
Beech..........................	43	11500	9360	....	9000 to 20000	1350000

Materials.	Weight of a cubic foot.	Ultimate resistance to—				Modulus of elasticity.
		Tearing.	Crushing.	Shearing.	Cross-br'k Modulus of Rupture.	
TIMBER—Continued.						
Birch	44	15000	6400	....	11700	1645000
Box	60	20000	10300			
Chestnut.	33.4	10000 to 13000	5300	....	1066	1140000
Elm	34	14000	10300	1400	6000 to 9[illegible]00	700000 to 1340000
Ebony, West Indian	74.5	.....	19000	....	2700	
Fir, Red Pine	37	12000 to 14000	5375 to 6200	500 to 800	7100 to 9540	146 000 to 1900000
" Spruce	37	12400	5900	600	9906 to 12300	1400[illegible]00 to 1800000
" Larch	33	9000 to 10000	5570	970 to 1700	50[illegible]0 to 10000	900000 to 1360000
Hickory	52	25000	11000	....	....	1040000
Hornbeam	47	20000	7300			
Lancewood	52.5	23400	....	....	17350	
Locust	44	16000	9000	....	11[illegible]00	
Lignum vitæ	62	11800	9900	....	1200	
Mahogany	35	8000 to 21800	8200	....	10000	1255000
Maple	49	10600	6500			
Oak, British	52.5	10000 to 19800	10000	2300	10000 to 13[illegible]00	1200000 to 1750000
" Dantzic	47.4	....	7700	....	8700	
" American white	42	18000	6100			
" " red	54	10250	6000	....	10609	2150000
Pine, American, white	34.6	11500	5300			
" yellow	29	15000	5400			
Sycamore	37	13000	12000	....	9600	
Teak, Indian	48	15000	12000	....	12000 to 19000	2400000
Water gum	62.5	....	11000	....	17460	
Walnut	40	8000	6500			
Willow, various	25	14000	4000	....	6600	
Yew	50	8000				
STONES, (natural and artificial.)						
Brick, weak red	125	......	550 to 800			
" strong red	135	280 to 300	1100			
" fire	137.5		1700			
" work	100	......	417 to 612			
Cement	89	280 to 300				

Materials.	Weight of a cubic foot.	Ultimate resistance to—				Modulus of elasticity.
		Tearing.	Crushing.	Shearing.	Cross-br'k. Modulus of Rupture.	
STONES—Continued.						
Chalk	145.5	118	330			
Glass	173	9400	....	....	....	8000000
Granite	168	....	5500 to 11000			
Limestone, marble	172	....	5500			
" granular	197	....	4000 to 4500			
Mortar, hydraulic	......	100 to 170				
" ordinary	109	50				
Rubble masonry	116	....	About 4-10 cut stone.			
Sandstone, strong	......	....	5500	....	2360	
" ordinary	144	....	3300 to 4400			
" weak	......	....	....	....	1100	
Slate	178	9600 to 12800	....	....	5000	13000000 to 16000000
MISCELLANEOUS.						
Flaxen yarn	......	25000				
Hempen ropes	......	14000				
Hide, ox, undressed	......	6300				
Leather, ox	......	4200				
Silk fibre	......	5200				
Whalebone	......	7700				

MODULUS OF RUPTURE R.

According to Professor Rankine, the modulus of rupture is eighteen times the weight that is required to break a bar of a given material one inch square (section) and one foot between supports, the weight concentrated at the middle.

MODULUS OF ELASTICITY E

Is that power (in lbs. generally) through which a prismatic body of a given material, of section $= 1$, is assumed to be extended double its length, or compressed to 0.

Let $A =$ Sectional area of a rod of the material.

$W =$ Weight or power in lbs., which causes the extension or compression of the rod.

$l =$ Length in inches of rod before W is applied.

$\gamma =$ The extension or compression of the rod in inches, caused by W.

Then will $E = \dfrac{Wl}{A l}$; $\gamma = \dfrac{W}{AE} \cdot l$.

FACTORS OF SAFETY k.

The ultimate resistance of material should be divided by—

Average, Steel and Wrought Iron.	For Proof Strength.	For Working Stress.
Steady load..................	2	3
Moving load.................	....	4 to 6
Cast Iron.		
Steady load..................	2 to 3	3 to 4
Moving load.................	....	6 to 8
Timber.		
Average......................	3	8 to 10

RESISTANCE TO CROSS-BREAKING AND SHEARING.

CAPACITY AND STRENGTH OF BEAMS.

Reference.

A = Area of cross-section of beam.
D = Deflection of beam from a horizontal.
E = Modulus of elasticity.
I = Moment of inertia of cross-section.
M = Maximum moment of rupture, or bending moment.
R = Modulus of rupture.
S = Vertical shearing force.
V = Pressure on supports.
W = Capacity or weight of load.
c, d, l = Dimensions in units of length.
k = Factor of safety.
w = Weight of load per unit of length.
$\frac{I}{s}$ = Moment of resistance of cross-section.

For the stability of a beam: $M = K = \frac{R}{k} \cdot \frac{I}{s}$.

The web of a metal beam must have sufficient area to resist the hearing force S; that is, $A = \frac{Sk}{\text{Ultimate resistance to shearing}}$.

The weight of the beam must be added to W, except in small eams, under 60 lbs. per lineal foot, when it may be disregarded.

[NOTE.—Always use the same units of dimensions or weight.]

No. of Figure.	Manner of loading and fastening beams.	Maximum moment of rupture M.	Capacity W of any section.
61	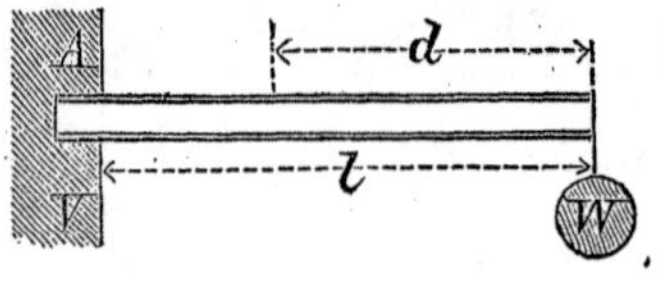	$W.l$	$\frac{K}{l}$
62	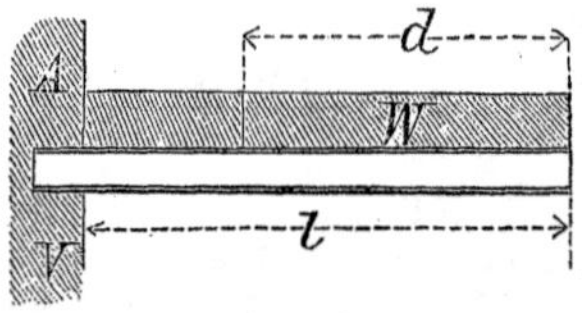	$W.\frac{l}{2}$	$2\frac{K}{l}$
63	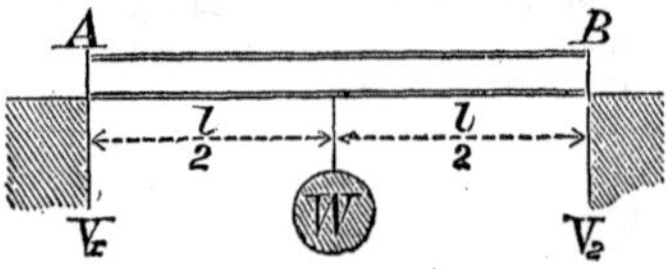	$W.\frac{l}{4}$	$4.\frac{K}{l}$
64	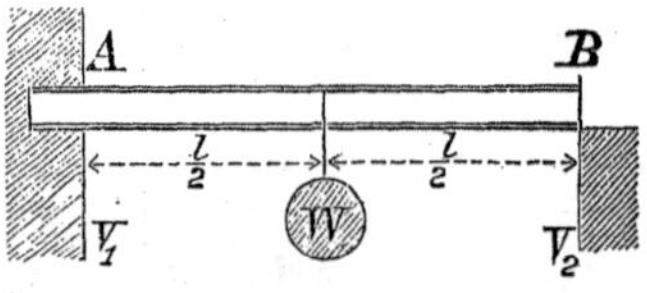	$W.\frac{l}{5.333}$	$5.333\frac{K}{l}$
65	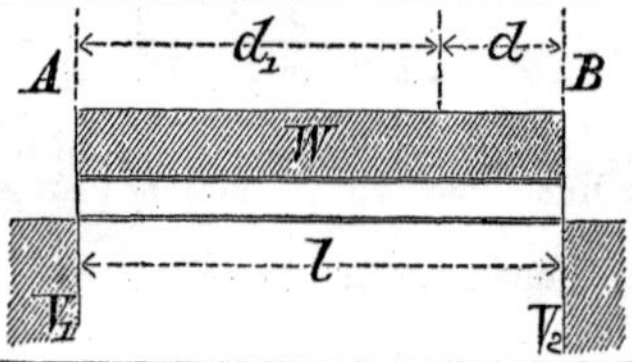	$W.\frac{l}{8}$	$8.\frac{K}{l}$

Maximum deflection D.	Distance from A to point of maximum D.	Shearing force S.	Pressure on supports V.
$\frac{W}{E.I}\cdot\frac{l^3}{3}$	l	At any point. W	W
$\frac{W}{E.I}\cdot\frac{l^3}{8}$	l	At any point. $w.d$	W
$\frac{W}{E.I}\cdot\frac{l^3}{48}$	$\frac{l}{2}$	At any point. $\frac{W}{2}$	$V_1 = V_2 = \frac{W}{2}$
$\frac{W.l^3}{E.I}.0.00931$	$0.553.l$	$\frac{3}{8}\ W.\frac{l}{2}$	$V_1 = V_2 = \frac{W}{2}$
$\frac{5}{8}\ \frac{W}{E.I}\cdot\frac{l^3}{48}$	$\frac{l}{2}$	At any point, $d < d_{\prime}$; $w\left(\frac{l}{2} - d\right)$	$V_1 = V_2 = \frac{W}{2}$

No. of Figure.	Manner of loading and fastening beams.	Maximum moment of rupture M.	Capacity W of any section.
66	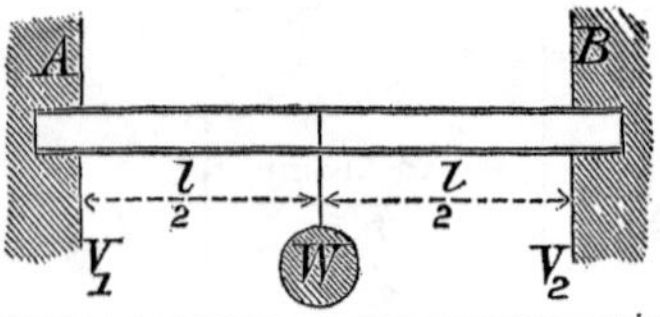	$W.\frac{l}{8}$	$8.\frac{K}{l}$
67	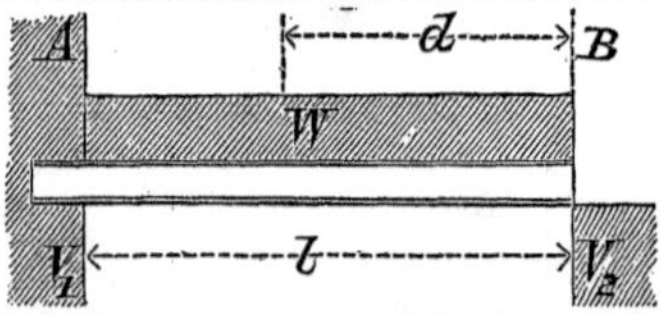	$W.\frac{l}{8}$	$8.\frac{K}{l}$
68	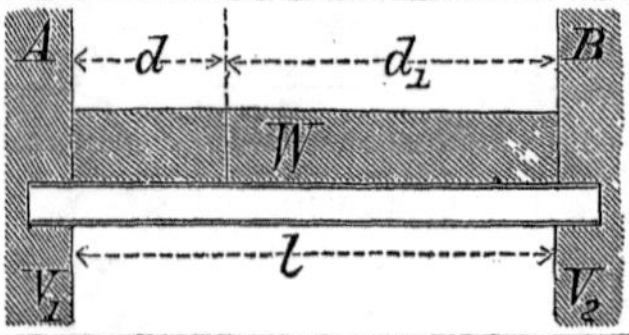	$W.\frac{l}{12}$	$12.\frac{K}{l}$
69	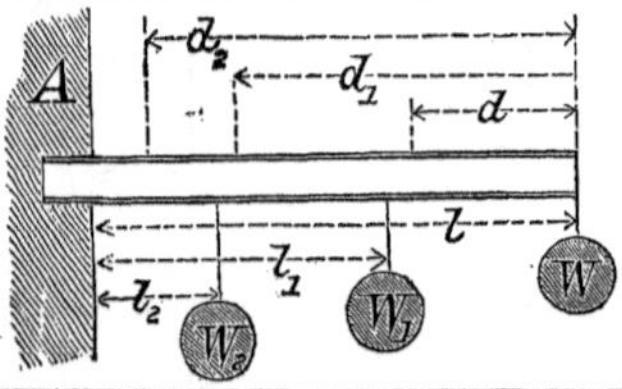	$W.l + W_1.l_1 + W_2.l_2$	
70	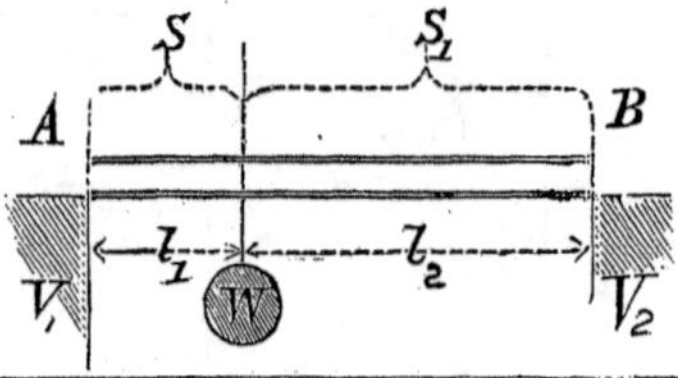	$W.\frac{l_1\,l_2}{l}$	$\frac{l}{l_1\,l_2}.K$

Maximum deflection D.	Distance from A to point of maximum D.	Shearing force S.	Pressure on supports V.
$\frac{W}{E.I}.\frac{l^3}{4.48}$	$\frac{l}{2}$	$\frac{W}{2}$	$V_1 = V_2 = \frac{W}{2}$
$\frac{W.l^3}{E.I}.0.0054$	$0.572.l$	$w.\left(\frac{3l}{8} - d\right)$	$V_1 = V_2 = \frac{W}{2}$
$\frac{W}{E.I}.\frac{l^3}{8.48}$	$\frac{l}{2}$	$d < d_{\prime}$; $w.\left(\frac{l}{2} - d\right)$	$V_1 = V_2 = \frac{W}{2}$
$\left(\frac{W_2}{E.I}.\frac{l_2^3}{3}\right) +$ $\left(\frac{W_1}{E.I}.\frac{l_1^3}{3}\right) +$ $\left(\frac{W}{E.I}.\frac{l^3}{3}\right)$	l	At any point between loads. $S = W$. $S_1 = W + W_1$. $S_2 = W + W_1 + W_2$	$W + W_1 + W_2$
$\frac{W}{E.I}.\frac{l^3}{3}\ \frac{l_2^2}{l^2}.\frac{l_1^2}{l^2}$		At any point and under any load. $S = W.\frac{l_2}{l}$ Constant bet. A & W. $S_1 = W.\frac{l_1}{l}$ Constant bet. B & W.	$V_1 = \frac{l_2}{l}\ W$ $V_2 = \frac{l_1}{l}\ W$

No. of Figure.	Manner of loading and fastening beams.	Maximum moment of rupture M.
71	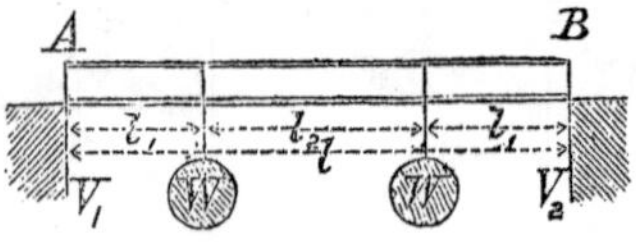	$W.l_1$
72	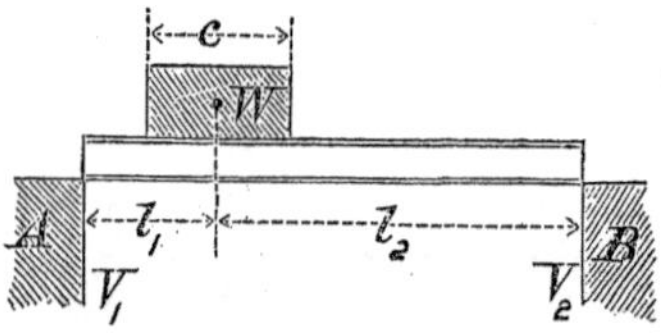	$W.\frac{l_1 l_2}{l}\left(1-\frac{c}{2l}\right)$
73	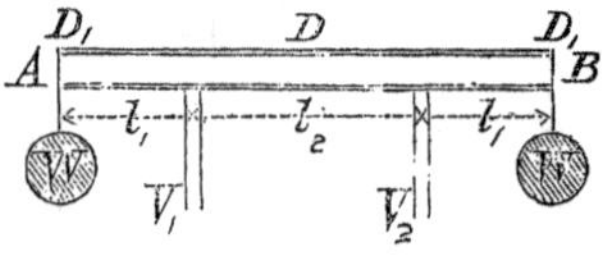	$W.l_1$
74	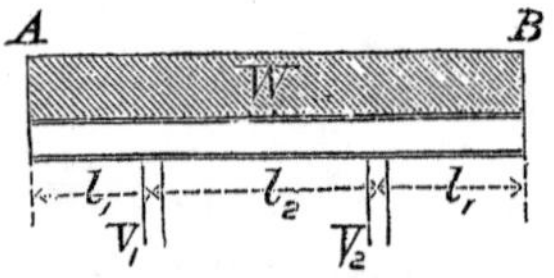	When $l > l_1\sqrt{8}$; $\frac{W}{2(l+2l_1)}.\left[\left(\frac{l}{2}\right)^2-l_1^2\right]$ When $l < l_1\sqrt{8}$; $\frac{Wl_1^2}{2(l+2l_1)}$

Capacity W of any section.	Maximum deflection D.	Distance from A to point of maximum D.	Shearing force S.	Pressure on supports V.
$\frac{K}{l_1}$	$\frac{W}{E.I} \cdot \frac{l_2{}^3}{8} \cdot \frac{l_1}{l_2}$	$\frac{l}{2}$	W	$V_1 = V_2 = W$
$\frac{Kl}{l_1 l_2 \left(1 - \frac{c}{2l}\right)}$			S at $A = W \frac{l_2}{l}$ S at $B = W \frac{l_1}{l}$	$V_1 = \frac{l_2}{l} W$ $V_2 = \frac{l_1}{l} W$
$\frac{K}{l_1}$	$D = \frac{W l_2{}^2 l_1}{8\,E.I}$ $D_1 = \frac{W l_1{}^2}{l} \cdot \left(\frac{l_2}{2} + \frac{l_1}{3}\right)$		W	$V_1 = V_2 = W$
$\frac{2\,(l + 2l_1)}{\left(\frac{l}{2}\right)^2 - l_1{}^2} K$			$w . l_1$ or $w . \frac{l_2}{2}$ The greater value to be taken.	$V_1 = V_2 = \frac{W}{2}$
$\frac{2\,(l + 2l_1)}{l_1{}^2} K$				

No. of Figure.	Manner of loading and fastening beams.	Maximum moment of rupture M.
75	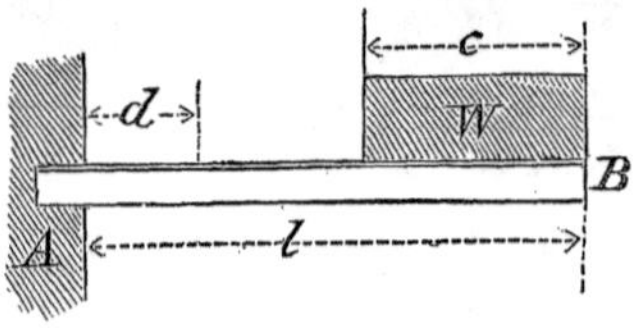 A, B, d, c, W, l	When $d < (l - c)$; $W.c\,(l - \frac{c}{2} - d)$ When $d > (l - c)$; $\frac{W}{2}(l - d)^2$
76	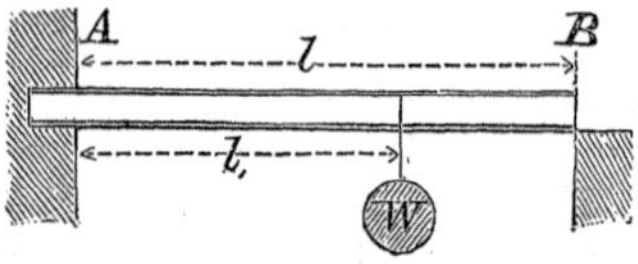 A, B, l, l_1, W	$W\frac{l_1^{\,2}(3l - l_1)(l - l_1)}{2l^3}$
77	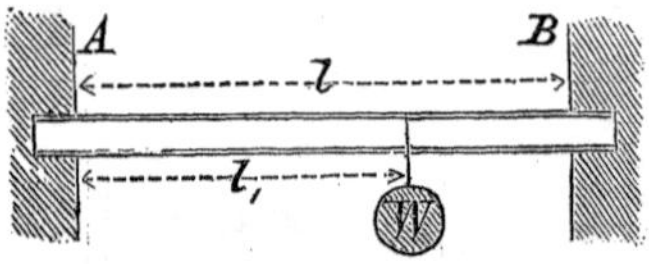 A, B, l, l_1, W	$W\frac{l_1^{\,2}(l - l_1)}{l^2}$
78	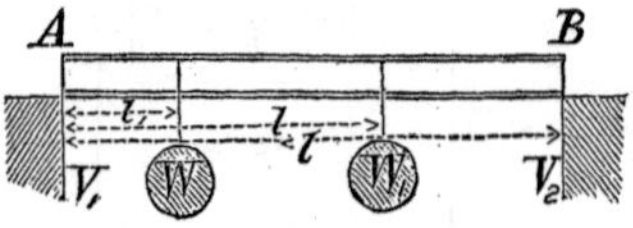 A, B, l_1, l_2, l, V_1, W, W_1, V_2	$\left[W\frac{l_1}{l}(2l - l_1 - l_2)\right]$ $\left[W_1\frac{l - l_2}{l}(l_1 + l_2)\right]$

Capacity W of any section.	Maximum deflection D.	Distance from A to point of maximum D.	Shearing force S.	Pressure on supports V.
$\frac{1}{c\,(l - \frac{c}{2} - d)} K$				
$\frac{2\,K}{(l - d)^2}$				
$\frac{2l^3}{l_1{}^2\,(3l - l_1)\,(l - l_1)} K$	$\frac{W}{3\,E.I} \cdot \frac{l_1{}^2\,(l - l_1)}{l^2}$			
$\frac{l^2}{l_1{}^2\,(l - l_1)} K$				

EXAMPLE.—Capacity of wrought-iron I-shaped beams; top and bottom flange alike; load equally distributed; ends not fixed.

Dimensions of Cross-section.

h = Height = 10 inches.
b = Width of flange = 4 inches.
t = Thickness of flange = 0.8 inches.
$t_{\prime}$ = Thickness of web = 0.5 inches.
$h_{\prime} = h - 2t;\ b_{\prime} = b - t_{\prime}$.

Distance between supports = 20 feet = 240 inches. Factor of safety = 3.

MOMENT OF RESISTANCE.

$$\frac{I}{s} = \frac{bh^3 - b_{\prime}h_{\prime}^3}{6h} = \frac{4 \times 10^3 - 3.5 \times 8.4^3}{6 \times 10} = 32.09.$$

Capacity W.

$$w = (4 \times 0.8 \times 2 + 8.4 \times 0.5) \times 240 \times 0.28 = 712.32 \text{ lbs.}$$

$$K = \frac{R}{k} \cdot \frac{I}{s} = \frac{38000}{3} \cdot 32.09 = 406473.33.$$

$$W = 8\frac{K}{l} - w = 8 \cdot \frac{406473.33}{240} - 712.32 = 12836.72 \text{ lbs.}$$

EXAMPLE.—Capacity of cast-iron ⊥-shaped beams; load equally distributed; ends not fixed; flange down.

Dimensions of Cross-section.

Let h = Height = 18 inches.
b = Width of flange = 9 inches.
t = Thickness of flange = 1.25 inches.
$t_{\prime}$ = Thickness of web = 1 inch.
$h_{\prime} = h - t;\ b_{\prime} = b - t_{\prime}$.

Area = 28 square inches. Distance between supports = 20 feet = 240 inches. Factor of safety k = 4.

MOMENT OF RESISTANCE.

$$\frac{I}{s} = \frac{1}{6}\left[\frac{(bh^2 - b_{\prime}h_{\prime}^2)^2}{bh^2 - 2b_{\prime}hh_{\prime} + b_{\prime}h_{\prime}^2} - \frac{4bhb_{\prime}h_{\prime}(h - h_{\prime})^2}{bh^2 - 2b_{\prime}hh_{\prime} + b_{\prime}h_{\prime}^2}\right]$$

$$= \frac{1}{6}\left[\frac{(9 \times 18^2 - 8 \times 16.75^2)^2}{9 \times 18^2 - 2 \times 8 \times 18 \times 16.75 + 8 \times 16.75^2} - \frac{4 \times 9 \times 18 \times 8 \times 16.75\ (18 - 16.75)^2}{9 \times 18^2 - 2 \times 8 \times 18 \times 16.75 + 8 \times 16.75^2}\right]$$

$$= \frac{1}{6}\left[\frac{452256.25}{336.5} - \frac{135675.00}{336\ 5}\right] = 157.$$

Capacity W.

$$w = 28 \times 240 \times 0.261 = 1754.\ \text{lbs.}$$

$$K = \frac{R}{k} \cdot \frac{I}{s} = \frac{28000}{4} \cdot 157 = 1099000.$$

$$W = 8\,\frac{K}{l} - w = 8 \cdot \frac{1000000}{240} - 1757 = 34879\ \text{lbs.}$$

For light beams no attention need be paid to weight of beam w.

CAPACITY W OF ROLLED I-SHAPED BEAMS.

Load equally distributed.

The calculations are based upon the patterns or sections used by the Phœnixville Iron Company. Practically this applies to all similar beams rolled in the United States, the difference in the profile of section being slight.

In the following table the factor of safety $k = 2.53$:

Reference.

$W =$ Load in tons of 2,000 lbs., equally distributed.
$w =$ Weight of beam in tons of 2,000 lbs.
$L =$ Distance between supports in feet.
$l =$ Distance between supports in inches.
$w_{\prime} =$ Weight per square foot of floor.
$W_{\prime} =$ Capacity of coupled or trebled beams in tons of 2,000 lbs.
$D =$ Deflection in inches at centre, between supports.
$d =$ Distance between centres of beams, when spacing for floors, in feet.

$$W = 8 \cdot \frac{K}{l} - w,\ K = \frac{R}{k} \cdot \frac{I}{s},\ \frac{R}{k} = \frac{38000}{2.53} = 15000\ \text{lbs.} =$$

$$7.5\ \text{tons.}\quad d = \frac{W}{L.w_{\prime}},\ \text{or}\ d = \frac{W_{\prime}}{L.w_{\prime}},\ D = \frac{5}{8}\,\frac{W + w}{E.I} \cdot \frac{l^3}{48}.$$

$K^1 =$ Constant, computed by formulas. (See under examples.)

The rivets for coupled or trebled beams should be about $\frac{3}{4}$ inch in diameter, and 8 inches apart.

Trebled Beams.

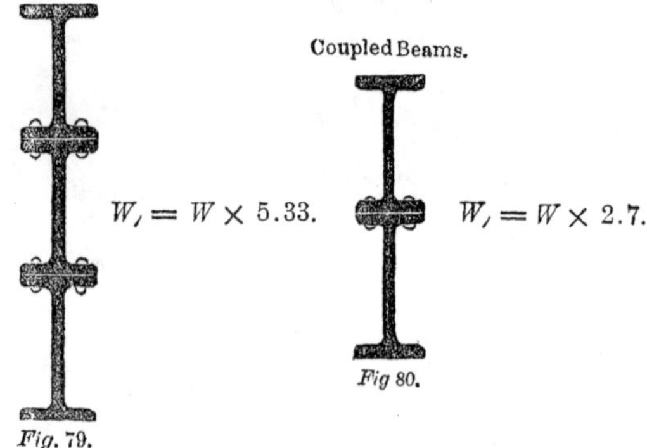

Fig. 79. *Fig* 80.

Examples explanatory of the following Table.

EXAMPLE.—What is the capacity of a 15-inch light beam, load equally distributed, distance between supports = 20 feet?

$K^1 = \frac{K \cdot 8}{12}$, and $W = \frac{K^1}{L}$; for 15-inch light beam $\frac{K^1}{L} =$

$\frac{345.19}{20} = 17.2$ *tons.* This is also found at the intersection of 20 feet and column under capacity W.

EXAMPLE.—What distance apart should 9-inch medium beams be placed, the distance between supports being 20 feet, and to carry a total load of 140 lbs. per square foot of floor surface?

Ans. 4.4 feet; being found at the intersection of the horizontal line from 20 feet and the vertical column under 140 lbs.

EXAMPLE.—What is the capacity of 12-inch light beams trebled, load equally distributed, distance between supports = 25 feet?

Ans. W for 12-inch light beam = 9.19 and $W_{\prime} = W \times 5.33 = 9.19 \times 5.33 = 48.98$ tons.

Capacity of Rolled Beams.

Explanation of Tables for I Beams.

The first column gives the distance between supports in feet.

The second column gives the capacity in tons of 2,000 lbs., equally distributed.

The third column gives the deflection in inches at centre of beam.

The fourth column gives the weight of beam in lbs. for length between supports.

The fifth to fifteenth column (inclusive) gives the distance in feet that the beams should be spaced from centre to centre, for weight in lbs., per sq. ft. of surface for floors.

Pounds in decimals of a ton.

lbs.		*tons.*
60	=	0.03
70	=	0.035
80	=	0.04
90	=	0.045
100	=	0.05
140	=	0.07
160	=	0.08
180	=	0.085
200	=	0 1
250	=	0.125
300	=	0.15

In using these beams for floors, with brick arching, the ends resting on supports should have a bearing of about 8 inches, resting on a cast-iron plate, 8 × 12 in. sq., by 1 in. thick.

Tie rods should be used where floors are subject to heavy concentrated moving loads, (as trucks with merchandise, &c.;) these rods should be about 8 times the depth of beam apart, fastened about ⅓ from the bottom of beam.

When beams are used to support walls, or as girders to carry floor beams, and put side by side (II,) they should be fastened together with cast-iron blocks, fitting between the flanges, so as to securely combine the two beams. The blocks may be put about the same distance apart as the tie-rods.

15″ "*Heavy*" *Beam*. Weight per lf. = 66.66 lbs.

Fig. 81.

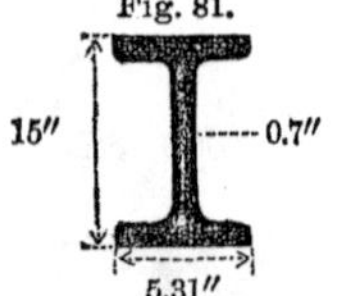

Sectional area......... = 20.0″
Moment of inertia I = 652.42
Constant K'............ = 434.95

$$W = \frac{K'}{L}.$$

Dis. bet. supports in feet.	Capac. W in tons.	Deflec. in inches.	Weight in lbs.	Distance d bet. centres of beams in feet, for weight in lbs. per sq. foot of—										
				60 lbs.	70 lbs.	80 lbs.	90 lbs.	100 lbs.	140 lbs.	160 lbs.	180 lbs.	200 lbs.	250 lbs.	300 lbs.
6	72.49	0.037	400.0											
7	62.13	0.050	466.6											
8	54.35	0.065	533 3											
9	48.32	0.084	600.0											
10	43.48	0.104	666.6											
11	39.54	0.126	733.3											
12	36.24	0.150	800.0	......	......	......	......	......	......	......	......	......	......	20.1
13	33.45	0.177	866.6	......	......	......	......	......	......	......	......	......	20.9	17.6
14	31.05	0.205	933.3	......	......	......	......	......	......	......	......	22.1	17.7	14.7
15	28.99	0.236	1000.0	......	......	......	......	......	......	......	22.3	19.3	15.5	12.8
16	27.18	0.270	1066.6	......	......	......	......	......	......	21.2	18.8	16.9	13.5	11.3
17	25.58	0 305	1133.3	......	......	......	......	......	21.4	19.6	17.0	15.0	12.0	10.2
18	24.16	0.342	1200.0	......	......	......	......	......	19.1	16.7	14.9	13.4	10.7	8.9
19	22.89	0.383	1266.6	......	......	......	......	......	17.6	15.2	13.4	12.0	9.6	8.0
20	21.73	0.426	1333.3	......	......	......	......	21.7	15.5	13.5	12.0	10.8	8.6	7.2
21	20.71	0.471	1400.0	......	......	......	22.0	19.7	14.7	12.5	11.5	9.8	7.9	6.7
22	19.58	0.515	1466.6	......	......	......	19.7	17.8	12 7	11.1	9.8	8.9	7.1	5.9
23	18.91	0.569	1533.3	......	......	21.0	18.9	17.1	11.8	10.5	9.4	8.2	6.7	5.5
24	18.12	0.623	1600.0	......	21.5	18.8	16.7	15.1	10.7	9.4	8.3	7.5	6.0	5.0
25	17.39	0.677	1666.6	......	19.9	17.3	15.5	14.4	10.2	8.6	7.7	6.9	5.6	4.7
26	16.72	0.735	1733.3	21.4	18.3	16.7	15.2	12.8	9.2	8.0	7.2	6.4	5.2	4.2
27	16.10	0.795	1800.0	19.8	17.1	14.9	13.4	11.9	8.5	7.4	6.7	5.9	4.7	3.9
28	15.53	0.860	1866.6	18.2	15.8	13.8	12.3	11.0	7.9	6.9	6.2	5.5	4.4	3.6
29	14.99	0.925	1933.3	17.2	14.8	12.9	11.5	10.7	7.4	6.8	5.7	5.1	4.1	3.4
30	14.49	0.994	2000.0	16.1	13.8	12.0	10.7	9.8	6.9	6.0	5.3	4.8	3.8	3.2
31	14.03	1.067	2066.6	15.0	12.9	11.3	10.0	9.0	6.4	5.6	5.0	4.5	3.6	3.0
32	13.59	1.141	2133.3	14.0	12.0	10.0	9.4	8.4	6.0	5.3	4.7	4.2	3.3	2.8
33	13.17	1.219	2200.0	13.3	11.4	9.9	8.8	7.9	5.7	4.9	4.4	3.9	3.1	2.6
34	12.79	1.304	2266.6	12.5	10.7	9.4	8.2	7.5	5.3	4.7	4.1	3.7	3.0	2.5
35	12.42	1.384	2333.3	11.8	10.1	8.8	7.9	7.1	5.0	4.4	3.9	3.5	2.8	2.3
36	12.08	1.473	2400.0	11.1	9.5	8.4	7.4	6.6	4.7	4.1	3 7	3.3	2.6	2.2
37	11.75	1.564	2466.6	10.8	9.1	7.9	7.0	6.3	4.4	3 9	3.5	3.1	2.5	2.1
38	11.43	1.656	2533.3	10.0	8.5	7.5	6.6	6.0	4.2	3.7	3.3	3.0	2.4	2.0
39	11.15	1.754	2600.0	9.5	8.1	7.1	6.3	5.7	4.0	3.5	3.1	2.8	2.2	1.9
40	10.87	1.854	2666.6	9.0	7.7	6.7	6.0	5.4	3.8	3.3	2.9	2.7	2.1	1.8

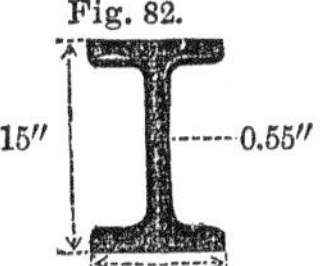

Fig. 82.

15″ "*Light*" *Beam.* Weight per lf. = 51.66 lbs.

Sectional area...... = 15.5″
Moment of inertia I = 517.78
Constant K'......... = 345.19

$$W = \frac{K'}{L}.$$

Dis. bet. supports in feet.	Capac. W in tons.	Deflec. in inches.	Weight in lbs.	Distance d bet. centres of beams in feet, for weight in lbs. per sq. foot of—										
				60 lbs.	70 lbs.	80 lbs.	90 lbs.	100 lbs.	140 lbs.	160 lbs.	180 lbs.	200 lbs.	250 lbs.	300 lbs.
6	57.52	0 037	310.0											
7	49.31	0.050	361.6											
8	43.13	0.065	413.3											
9	38.35	0.084	465.0											
10	34.50	0.103	516.6	......	......	......	......	......	......	......	......	......	......	23.0
11	31.38	0.124	567.3	......	......	......	......	......	......	......	......	......	22.9	19.0
12	28.76	0.150	620.0	......	......	......	......	......	......	......	......	......	19.1	15.9
13	26.55	0.176	671.6	......	......	......	......	......	......		22.7	20.4	16.4	13.6
14	24.65	0.205	723.3	......	......	......	......	......	......	22.0	19.5	17.6	14.0	11.7
15	23.01	0.236	775.0	......	......	......	......	......	21.9	19.1	17.0	15.3	12.3	10.2
16	21.57	0.269	806.6	......	......	......	......	......	19.2	16.8	14.9	13.4	10.7	8.9
17	20.30	0.304	858.3	......	......	......	......	......	17.9	14.9	13.2	11.9	9.5	7.9
18	19.16	0.341	930.0	......	......	......	......	21.3	15.2	13.3	11.8	10.6	8.5	7.1
19	18.15	0.381	981.6	......	......	......	21.3	19.1	13.6	11.9	10.6	9.5	7.6	6.3
20	17.24	0.424	1033.3	......	......	21.5	19.1	17.2	12.3	10.7	9.5	8.6	6.9	5.7
21	16.43	0.469	1085.0	......	......	19.5	17.4	15.6	11.1	9.8	8.7	7.8	6.2	5.2
22	15.68	0.515	1136.6	......	20.3	17.8	15.8	14.2	10.1	8.9	7.9	7.1	5.7	4.7
23	15.00	0.565	1187.3	21.7	18.7	16.3	14.5	13.0	9.3	8.1	7.2	6.5	5.2	4.3
24	14.38	0.620	1240.0	19.9	17.1	14.9	13.3	11.9	8.5	7.4	6.6	5.9	4.8	3.9
25	13.80	0.674	1291.6	18.4	15.8	13.8	12.3	11.0	7.8	6.9	6.1	5.5	4.4	3.6
26	13.27	0.732	1343.3	17.0	14.5	12.8	11.3	10.2	7.2	6.3	5.6	5.1	4.0	3.4
27	12.78	0.791	1395.0	15.7	13.6	11.8	10.5	9.4	6.7	5.9	5.2	4.7	3.7	3.1
28	12.32	0.855	1446.6	14.6	12.5	11.0	9.7	8.8	6.2	5.5	4.8	4.4	3.5	2.9
29	11.93	0.921	1498.3	13.7	11.8	10.2	9.1	8.2	5.8	5.1	4.5	4.1	3.2	2.7
30	11.50	0.989	1550.0	12.7	10.9	9.5	8.5	7.6	5.4	4.7	4.2	3.8	3.0	2.5
31	11.13	1.060	1601.6	11.9	10.3	8.9	8.0	7.1	5.1	4.4	3.9	3.5	2.8	2.3
32	10.78	1.133	1653.3	11.2	9.6	8.4	7.4	6.7	4.8	4.2	3.7	3.3	2.6	2.2
33	10.46	1.211	1705.0	10.5	9.0	7.9	7.0	6.3	4.5	3.9	3.5	3.1	2.5	2.1
34	10.14	1.292	1756.6	9.9	8.5	7.4	6.6	5.9	4.2	3.7	3.3	2.9	2.3	1.9
35	9.86	1.375	1808.3	9.3	8.0	7.0	6.2	5.6	4.0	3.5	3.1	2.8	2.2	1.8
36	9.58	1.463	1860.0	8.8	7.6	6.6	5.9	5.3	3.8	3.3	2.9	2.6	2.1	1.7
37	9.32	1.553	1911.6	8.3	7.2	6.2	5.6	5.0	3.5	3.1	2.7	2.5	2.0	1.6
38	9.08	1.645	1963.3	7.9	6.8	5.9	5.3	4.7	3.4	2.9	2.6	2.3	1.9	1.5
39	8.85	1.742	2015.0	7.5	6.5	5.6	5.0	4.5	3.2	2.8	2.5	2.2	1.8	1.4
40	8.62	1.841	2066.6	7.1	6.1	5.3	4.7	4.3	3.0	2.6	2.3	2.1	1.7	1.4

12″ "*Heavy*" *Beam.* Weight per lf. = 56.66 lbs.

Fig. 83.

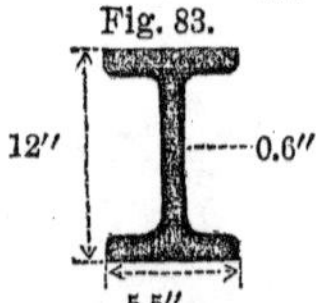

Sectional area...... = 17.0″
Moment of inertia I = 373.53
Constant K'........ = 311.28

$$W = \frac{K'}{L}.$$

Dis. bet. supports in feet.	Capac. W in tons.	Deflec. in inches.	Weight in lbs.	Distance d bet. centres of beams in feet, for weight in lbs. per sq. foot of—										
				60 lbs.	70 lbs.	80 lbs.	90 lbs.	100 lbs.	140 lbs.	160 lbs.	180 lbs.	200 lbs.	250 lbs.	300 lbs.
6	51.88	0.046	340.0											
7	44.54	0.063	396.6											
8	38.70	0.082	453.3											
9	34.58	0.105	510.0											
10	31.12	0.131	566.6	......	......	......	......	......	......	......	......	......	......	20.7
11	28.29	0.158	623.3	.....	......	......	......	......	......	......	......	......	20.5	17.1
12	25.94	0.188	680.0	......	......	.....	......	......	......	......	......	21.6	17.2	14.4
13	23.94	0.222	736.6	......	.....	......	....	......	......	23.0	20.4	18.4	14.7	12.2
14	22.22	0.258	793.3	......	......	.....	......	......	22.6	19.8	17.6	15.8	12.6	10.5
15	20.75	0.297	850.0	......	......	......	......	......	19.7	17.2	15.3	13.8	11.0	9.2
16	19.50	0.339	906.6	......	......	......	......	....	17.4	15.2	13.5	12.1	9.7	8.1
17	18.31	0.383	963.3	......	......	......	......	21.5	15.3	13.4	11.9	10.7	8.6	7.1
18	17.29	0.431	1020.0	......	......	......	21.3	19.2	13.7	12.0	10.6	9.6	7.6	6.4
19	16.38	0.481	1076.6	......	....	21.5	19.1	17.2	12.3	10.7	9.5	8.6	6.8	5.7
20	15.61	0.538	1133.3	...	......	19.5	17.3	15.6	11.1	9.7	8.6	7.8	6.2	5.2
21	14.82	0.592	1190.0	......	20.1	17.6	15.6	14.1	10.0	8.8	7.8	7.0	5.6	4.7
22	14.14	0.652	1246.6	21.4	18.3	16.0	14 2	12.8	9.1	8.0	7.1	6.4	5.1	4.2
23	13.53	0.717	1303.3	19.6	16.8	14.7	13 0	11.7	8.4	7.3	6.5	5.8	4.7	3.9
24	12.97	0.786	1360.0	18.0	15.4	13.5	12.0	10.8	7.7	6.7	6.0	5.4	4.3	3.6
25	12.45	0.855	1416.6	16.6	14.2	12.4	11.0	9.9	7.1	6.2	5.5	4.9	3.9	3.3
26	11.97	0.927	1473.3	15.3	13.1	11.5	10.2	9.2	6.5	5.7	5.1	4.6	3.6	3.0
27	11.52	1.003	1530.0	14.2	12.1	10.6	9.4	8.5	6.0	5.3	4.7	4.2	3.4	2.8
28	11.11	1.084	1586.6	13.2	11.3	9.9	8.8	7.9	5.6	4.9	4.4	3.9	3.1	2.6
29	10.73	1.170	1643.8	12.3	10.5	9.2	8.2	7.4	5.2	4.6	4.1	3.7	2.9	2.4
30	10.37	1.257	1700.0	11.5	9.8	8.6	7.6	6.9	4.9	4.3	3.8	3.4	2.7	2.3
31	10.04	1.350	1756.6	10.	9.2	8.0	7.1	6.4	4.6	4.0	3.6	3.2	2.5	2.1
32	9.71	1.443	1813.3	10.1	8.6	7.5	6.7	6.0	4.3	3.7	3.4	3.0	2.4	2.0
33	9.43	1.546	1870.0	9.5	8.2	7.1	6.3	5.7	4.0	3.5	3.1	2.8	2.2	1.9
34	9.15	1.650	1926.6	8.9	7.6	6.7	5.9	5.3	3.8	3.3	2.9	2.6	2.1	1.7
35	8.89	1.758	1983.3	8.4	7.2	6.3	5.6	5.0	3.6	3.1	2.8	2.5	2.0	1.6
36	8.64	1.871	2040.0	8.0	6.8	6.0	5.3	4.8	3.4	3.0	2.6	2.4	1.9	1.6
37	8.41	1.987	2096.6	7.5	6.4	5.6	5.0	4.5	3.2	2.8	2.5	2.2	1.8	1.5
38	8.18	2.104	2153.3	7.1	6.1	5.3	4.7	4.3	3.0	2.6	2.3	2.1	1.7	1.4
39	7.98	2.234	2210.0	6.8	5.8	5.1	4.5	4.0	2.9	2.5	2.2	2.0	1.6	1.3
40	7.78	2.336	2266.6	6.4	5.5	4.8	4.3	3.8	2.7	2.4	2.1	1.9	1.5	1.2

12″ "*Light*" *Beam.* Weight per lf. = 41.66 lbs.

Fig. 84.

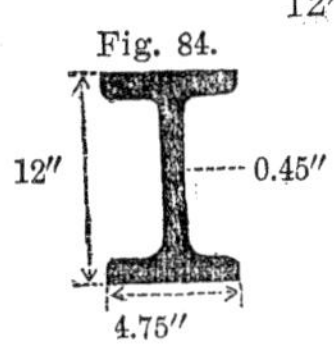

Sectional area......... = 12.5″
Moment of inertia I = 275.92
Constant K'............. = 229.94

$$W = \frac{K'}{L}.$$

Dis. bet. supports in feet.	Capac. W in tons.	Deflec. in in.	Weight in lbs.	Distance d bet. centres of beams in feet, for weight in lbs. per sq. foot of—										
				60 lbs.	70 lbs.	80 lbs.	90 lbs.	100 lbs.	140 lbs.	160 lbs.	180 lbs.	200 lbs.	250 lbs.	300 lbs.
6	39.31	0.047	250.0											
7	32.84	0.063	291.6											
8	28.74	0.083	333.3	......	......	......	......	......	......	......	......	......	......	24.0
9	25.54	0.105	375.0	......	......	......	......	......	......	......	......	......	23.0	18.9
10	22.98	0.131	416.6	......	......	......	......	......	......	......	......	22.0	18.3	15.3
11	20.90	0.158	458.3	......	......	......	......	......	......	23.0	21.0	19.0	15.2	12.6
12	19.16	0.180	500.0	......	......	......	......	......	22.0	19.9	17.7	15.9	12.7	10.6
13	17.68	0.222	541.6	......	......	......	......	......	19.4	17.0	15.1	13.6	10.9	9.0
14	16.42	0 258	583.3	......	......	......	......	......	16.7	14.6	13.0	11.7	9.3	7.8
15	15.32	0.297	625.0	......	......	......	22.0	20.0	14.5	12.7	11.3	10.2	8.1	6.7
16	14.37	0.339	666.6	......	......	22.0	19.9	17.9	12.8	11.2	9.9	8.9	7.1	5.9
17	13.52	0.383	708.3	......	......	19.9	17.7	15.9	11.3	9.9	8.8	7.9	6.3	5.3
18	12.77	0.431	750.0	......	20.0	17.7	15.7	14.1	10.1	8.8	7.8	7.1	5.6	4.7
19	12.10	0.481	791.6	21.0	18.3	15.9	14.2	12.7	9.1	7.9	7.0	6.3	5.1	4.2
20	11.48	0.538	833.3	19.1	16.4	14.3	12.7	11.4	8.2	7.1	6.3	5.7	4.5	3.8
21	10.94	0.592	875.0	17.3	15.0	13.0	11.6	10.4	7.4	6.5	5.7	5.2	4.1	3.4
22	10.44	0.652	916.6	15.8	13.5	11.8	10.5	9.5	6.7	5.9	5.2	4.7	3.7	3.1
23	9.99	0.717	958.3	14.4	12.5	10.8	9.7	8.6	6.2	5.4	4.8	4.3	3.4	2.8
24	9.58	0.786	1000.0	13.3	11.4	9.9	8.8	7.9	5.7	4.9	4.4	3.9	3.1	2.6
25	9.19	0.855	1041.6	12.2	10.5	9.1	8.2	7.3	5.2	4.5	4.0	3.6	2.9	2.4
26	8.84	0.927	1083.3	11.3	9.7	8.5	7.5	6.8	4.8	4.2	3.7	3.4	2.7	2.2
27	8.51	1.003	1125.0	10.5	9.0	7.8	7.0	6.3	4.5	3.9	3.5	3.1	2.5	2.1
28	8.21	1.084	1166.6	9.7	8.3	7.3	6.5	5.8	4.1	3.6	3.2	2.9	2.3	1.9
29	7.92	1.170	1208.3	9.1	7.8	6.8	6.1	5.4	3.8	3.4	3.0	2.7	2.1	1.8
30	7.66	1.257	1250.0	8.5	7.2	6.3	5.6	5.1	3.6	3.1	2.8	2.5	2.0	1.7
31	7.41	1.350	1291.6	7.9	6.8	5.9	5.3	4.8	3.4	2.9	2.6	2.3	1.9	1.6
32	7.18	1.443	1333.3	7.4	6.4	5.6	4.9	4.4	3.2	2.8	2.4	2.2	1.7	1.4
33	6.96	1.542	1375.0	7.0	6.0	5.2	4.7	4.2	3.0	2.6	2.3	2.1	1.6	1.4
34	6.75	1.645	1416.6	6.6	5.6	4.9	4.4	3.9	2.8	2.4	2.2	2.0	1.5	1.3
35	6.57	1.754	1458.3	6.2	5.3	4.7	4.1	3.7	2.6	2.3	2.0	1.8	1.5	1.
36	6.38	1.871	1500.0	5.9	5.0	4.4	3.9	3.5	2.5	2.2	1.9	1.7	1.4	1.
37	6.21	1.987	1541.6	5.5	4.8	4.2	3.7	3.3	2.3	2.0	1.8	1.6	1.3	1.
38	6.05	2.109	1583.3	5.3	4.5	3.9	3.5	3.1	2.2	1.9	1.7	1.5	1.2	1.
39	5.89	2.229	1625.0	5.0	4.3	3.7	3.3	3.0	2.1	1.8	1.6	1.4	1.1	1.
40	5.74	2.366	1666.6	4.7	4.1	3.5	3.1	2.8	2.0	1.7	1.5	1.3	1.0	0.

10.5″ *Beam.* Weight per lf. = 35 lbs.

Fig. 85.

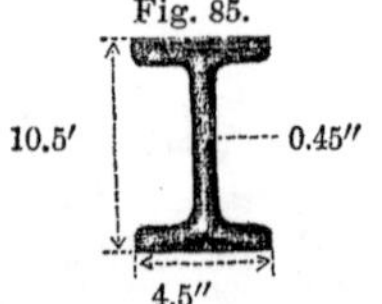

Sectional area......... = 10.5″
Moment of inertia I = 179.44
Constant K'............ = 170.903

$$W = \frac{K'}{L}.$$

Dis. bet. supports in feet.	Capac. W in tons.	Deflec. in in.	Weight in lbs.	Distance d bet. centres of beams in feet, for weight in lbs. per sq. foot of—										
				60 lbs.	70 lbs.	80 lbs.	90 lbs.	100 lbs.	140 lbs.	160 lbs.	180 lbs.	200 lbs.	250 lbs.	300 lbs.
6	23.48	0.053	210.0											
7	24.41	0.072	245.0	......	......	......	.....	......	......	......	......	......	......	23.2
8	21.36	0.095	280.0	......	......	......	......	......	......	......	......	......	21.3	17.8
9	18.98	0.120	315.0	......	......	......	......	......	......	......	23.4	21.1	17.0	14.0
10	17.09	0.149	350.0	......	......	......	......	......	......	21.3	18.9	17.0	13.6	11.4
11	15.53	0.181	385.0	......	......	......	......	......	20.1	17.6	15.6	14.1	11.3	9.3
12	14.24	0.216	420.0	......	......	......	......	......	16.9	14.8	13.1	11.8	9.4	7.9
13	13.14	0.254	455.0	......	....	......	20.6	20.2	14.4	12.6	11.1	10.1	8.1	6.7
14	12.20	0.295	490.0	......	......	21.7	19.2	17.4	12.4	10.9	9.6	8.7	6.9	5.8
15	11.38	0.340	525.0	....	21.9	18.9	17.0	15.1	10.8	9.4	8.4	7.5	6.0	5.0
16	10.68	0.389	560.0	22.2	19.0	16.6	14.9	13.3	9.5	8.3	7.4	6.6	5.3	4.4
17	10.05	0.439	595.0	19.7	17.0	14.7	13.2	11.8	8.4	7.3	6.5	5.9	4.7	3.9
18	9.49	0.494	630.0	17.5	15.0	13.1	11.7	10.5	7.6	6.5	5.8	5.2	4.2	3.5
19	8.99	0.553	665.0	15.7	13.6	11.7	10.5	9.4	6.7	5.9	5.2	4.7	3.7	3.1
20	8.54	0.614	700.0	14.2	12.2	10.6	9.4	8.5	6.1	5.3	4.7	4.2	3.4	2.8
21	8.13	0.681	735.0	12.9	11.1	9.6	8.6	7.7	5.5	4.8	4.3	3.8	3.1	2.5
22	7.75	0.752	770.0	11.7	10.0	9.1	7.8	7.0	5.0	4.4	3.9	3.5	2.8	2.3
23	7.43	0.823	805.0	10.7	9.2	8.0	7.2	6.4	4.6	4.0	3.5	3.2	2.5	2.1
24	7.12	0.903	840.0	9.8	8.4	7.4	6.5	5.9	4.2	3.7	3.2	2.9	2.3	1.9
25	6.83	0.980	875.0	9.1	7.8	6.8	6.0	5.4	3.9	3.4	3.0	2.7	2.1	1.8
26	6.57	1.067	910.0	8.4	7.2	6.3	5.6	5.0	3.6	3.1	2.8	2.5	2.0	1.6
27	6.32	1.154	945.0	7.8	6.7	5.8	5.2	4.6	3.3	2.9	2.6	2.3	1.8	1.5
28	6.10	1.251	980.0	7.2	6.2	5.4	4.8	4.3	3.1	2.7	2.4	2.1	1.7	1.4
29	5.89	1.346	1015.0	6.7	5.8	5.0	4.5	4.0	2.9	2.5	2.2	2.0	1.6	1.3
30	5.69	1.450	1050.0	6.3	5.4	4.7	4.2	3.7	2.7	2.3	2.1	1.8	1.5	1.2
31	5.51	1.556	1085.0	5.9	5.1	4.4	3.9	3.5	2.5	2.2	1.9	1.7	1.4	1.1
32	5.34	1.672	1120.0	5.5	4.7	4.1	3.7	3.3	2.3	2.0	1.8	1.6	1.3	1.1
33	5.17	1.783	1155.0	5.2	4.4	3.9	3.4	3.1	2.2	1.9	1.7	1.5	1.2	1.0
34	5.02	1.906	1190.0	4.8	4.2	3.6	3.2	2.9	2.1	1.8	1.6	1.4	1.1	
35	4.88	2.033	1225.0	4.6	4.0	3.4	3.1	2.7	1.9	1.7	1.5	1.3	1.1	
36	4.69	2.143	1260.0	4.3	3.7	3.2	2.8	2.6	1.8	1.6	1.4	1.3	1.0	
37	4.61	2.297	1295.0	4.1	3.5	3.1	2.7	2.4	1.7	1.5	1.3	1.2		
38	4.50	2.444	1330.0	3.9	3.3	2.9	2.6	2.3	1.6	1.4	1.3	1.1		
39	4.38	2.589	1365.0	3.6	3.2	2.8	2.5	2.2	1.6	1.4	1.2	1.1		
40	4.20	2.711	1400.0	3.5	3.0	2.6	2.3	2.1	1.5	1.3	1.1	1.0		

Fig. 86.

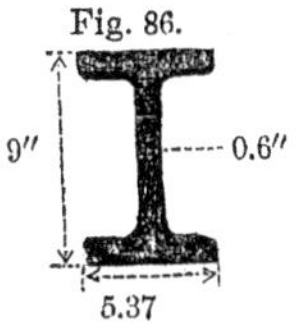

9″ 0.6″ 5.37

9″ "*Heavy*" *Beam.* Weight per lf. = 50 lbs.

Sectional area......... = 15.0″
Moment of inertia I = 188.55
Constant K'............ = 209.50

$$W = \frac{K'}{L}.$$

Dis. bet. supports in feet.	Capac. W in tons.	Deflec. in inches.	Weight in lbs.	Distance d bet. centres of beams in feet, for weight in lbs. per sq. foot of—										
				60 lbs.	70 lbs.	80 lbs.	90 lbs.	100 lbs.	140 lbs.	160 lbs.	180 lbs.	200 lbs.	250 lbs.	300 lbs.
6	36.91	0.065	300.0											
7	29.92	0.084	350.0											
8	26.18	0.111	400.0	......	......	......	......	......	......	......	......	......	......	21.8
9	23.27	0.141	450.0	......	......	......	......	......	......	......	......	......	20.6	17.2
10	20.95	0.174	500.0	......	......	......	......	......	......	......	23.2	20.9	16.7	13.9
11	19.04	0.211	550.0	......	......	......	......	......	......	21.6	19.1	17.3	13.7	11.5
12	17.45	0.253	600.0	......	......	......	......	......	20.7	18.1	16.1	14.5	11.6	9.6
13	16.11	0 297	650.0	......	......	......	......	......	17.6	15.4	13.8	12.3	9.9	8.2
14	14.96	0.345	700.0	......	......	......	......	21.3	15.2	13.3	11.8	10.6	8.5	7.1
15	13.96	0.398	750.0	......	......	......	20.6	18.6	13.2	11.6	10.3	9.3	7.5	6.2
16	13.09	0.454	800.0	......	......	20.4	19.5	16.3	11.6	10.2	9.0	8.1	6.5	5.4
17	12.32	0.515	850.0	......	20.7	18.1	16.1	14.5	10.3	9.0	8.0	7.2	5.7	4.8
18	11.63	0.580	900.0	21.5	18.4	16.1	14.3	12.9	9.2	8.0	7.1	6.4	5.1	4.3
19	11.02	0.648	950.0	19.3	16.5	14.5	12.8	11.6	8.2	7.2	6.4	5.8	4.6	3.8
20	10.47	0.722	1000.0	17.4	14.9	13.0	11.4	10.4	7.4	6.5	5.8	5.2	4.1	3.4
21	9.97	0.799	1050.0	15.8	13.5	11.8	10.5	9.4	6.8	5.9	5.2	4.7	3.4	3.1
22	9.52	0.883	1100.0	14.4	12.3	10.3	9.6	8.2	6.1	5.4	4.8	4.3	3.4	2.8
23	9.10	0.968	1150.0	13.1	11.3	9.8	8.7	7.9	5.6	4.9	4.3	3.9	3.1	2.6
24	8.72	1.062	1200.0	12.1	10.3	9.0	8.0	7.2	5.1	4.5	4.0	3.6	2.9	2.4
25	8.34	1.152	1250.0	11.1	9.5	8.3	7.4	6.6	4.7	4.1	3.7	3.3	2.6	2.2
26	8.05	1.258	1300.0	10.3	8.8	7.7	6.8	6.1	4.4	3.8	3.4	3.0	2.4	2.0
27	7.75	1.363	1350.0	9.5	8.0	7.1	6.3	5.7	4.1	3.5	3.1	2.8	2.2	1.9
28	7.48	1.477	1400.0	8.9	7.6	6.6	5.9	5.3	3.8	3.3	2.9	2.6	2.1	1.7
29	7.22	1.593	1450.0	8.3	7.2	6.2	5.5	4.9	3.5	3.1	2.7	2.5	1.9	1.6
30	6.98	1.718	1500.0	7.7	6.6	5.8	5.1	4.6	3.3	2.9	2.5	2.3	1.8	1.5
31	6.75	1.846	1550.0	7.3	6.2	5.4	4.8	4.3	3.1	2.7	2.4	2.1	1.7	1.4
32	6.54	1.982	1600.0	6.7	5.8	5.1	4.4	4.0	2.9	2.5	2.2	2.0	1.6	1.3
33	6.34	2.119	1650.0	6.4	5.6	4.8	4.2	3.8	2.7	2.4	2.1	1.9	1.5	1.2
34	6.16	2.265	1700.0	6.0	5.1	4.5	4.0	3.6	2.5	2.2	2.0	1.8	1.4	1.2
35	5.98	2.416	1750.0	5.6	4.8	4.2	3.7	3.4	2.4	2.1	1.8	1.7	1.3	1.1
36	5.81	2.577	1800.0	5.3	4.6	4.0	3.5	3.3	2.3	2.0	1.7	1.6	1.2	1.0
37	5.66	2.742	1850.0	5.0	4.3	3.8	3.3	3.0	2.1	1.9	1.6	1.5	1.2	
38	5.51	2.918	1900.0	4.8	4.1	3.6	3.2	2.9	2.0	1.8	1.6	1.4	1.1	
39	5.37	3.098	1950.0	4.5	3.9	3.4	3.0	2.7	1.9	1.7	1.5	1.3	1.1	
40	5.27	3.289	2000.0	4.3	3.7	3.2	2.9	2.6	1.8	1.6	1.4	1.3	1.0	

9″ "*Medium*" *Beam.* Weight per lf. = 30 lbs.

Fig. 87.

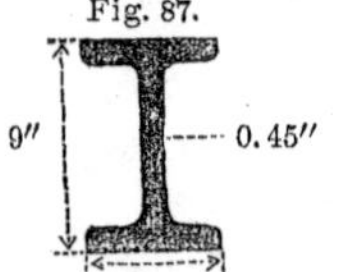

Sectional area.......... = 9.0″
Moment of inertia I = 111.32
Constant K'............ = 123.69

$$W = \frac{K'}{L}.$$

Dis. bet. supports in feet.	Capac. W in tons.	Deflec. in inches.	Weight in lbs.	Distance d bet. centres of beams in feet, for weight in lbs. per sq. foot of—										
				60 lbs.	70 lbs.	80 lbs.	90 lbs.	100 lbs.	140 lbs.	160 lbs.	180 lbs.	200 lbs.	250 lbs.	300 lbs.
6	20.60	0.062	180.0	......	......	......	......	......	......	......	......	......	......	22.0
7	17.67	0.085	210.0	......	......	......	......	......	......	......	......	25.0	20.0	16.0
8	15.46	0.111	240.0	......	......	......	......	......	......	24.0	21.0	19.0	15.0	12.0
9	13.74	0.141	270.0	......	......	......	......	......	21.0	19.0	16.0	15.0	11.0	10.0
10	12.36	0.174	300.0	......	......	......	......	......	17.0	15.0	13.0	12.0	9.8	8.2
11	11.24	0.211	330.0	......	......	......	22.0	20.0	14.0	12.0	11.0	10.0	8.1	6.8
12	10.30	0.252	360.0	......	......	21.0	19.0	17.0	12.0	10.0	9.5	8.5	6.8	5.7
13	9.51	0.297	390.0	......	20.0	18.0	16.0	14.0	10.0	9.1	8.1	7.3	5.8	4.8
14	8.83	0.345	420.0	21.0	18.0	15.0	14.0	12.0	9.0	7.8	7.0	6.3	5.0	4.2
15	8.24	0.398	450.0	18.0	15.0	13.0	12.0	10.0	7.8	6.8	6.1	5.4	4.3	3.6
16	7.73	0.455	480.0	16.0	13.0	12.0	10.0	9.6	6.9	6.0	5.3	4.8	3.8	3.2
17	7.21	0.511	510.0	14.0	12.0	10.0	9.4	8.4	6.0	5.3	4.7	4.2	3.2	2.8
18	6.87	0.580	540.0	12.0	10.0	9.5	8.4	7.6	5.4	4.7	4.2	3.8	3.0	2.5
19	6.51	0.650	570.0	11.0	9.7	8.5	7.6	6.8	4.8	4.2	3.8	3.4	2.7	2.2
20	6.18	0.722	600.0	10.0	8.8	7.7	6.8	6.1	4.4	3.8	3.4	3.0	2.4	2.0
21	5.88	0.799	630.0	9.3	8.0	7.0	6.2	5.6	4.0	3.5	3.1	2.8	2.2	1.8
22	5.62	0.884	660.0	8.5	7.2	6.3	5.6	5.1	3.6	3.1	2.8	2.5	2.0	1.7
23	5.37	0.969	690.0	7.7	6.6	5.8	5.1	4.6	3.3	2.9	2.5	2.3	1.8	1.5
24	5.15	1.065	720.0	7.1	6.1	5.3	4.7	4.2	3.0	2.6	2.3	2.1	1.7	1.4
25	4.94	1.157	750.0	6.5	5.6	4.9	4.3	3.9	2.8	2.5	2.1	1.9	1.5	1.3
26	4.83	1.277	780.0	6.1	5.3	4.6	4.1	3.7	2.6	2.3	2.0	1.8	1.4	1.2
27	4.58	1.365	810.0	5.6	4.8	4.2	3.7	3.3	2.4	2.1	1.8	1.6	1.3	1.1
28	4.41	1.476	840.0	5.2	4.5	3.9	3.5	3.1	2.2	1.9	1.7	1.5	1.2	1.0
29	4.26	1.593	870.0	4.8	4.1	3.6	3.2	2.9	2.0	1.8	1.6	1.4	1.1	
30	4.12	1.718	900.0	4.5	3.9	3.4	3.0	2.7	1.9	1.7	1.5	1.3	1.0	
31	3.99	1.846	930.0	4.2	3.6	3.2	2.8	2.5	1.8	1.6	1.4	1.2	1.0	
32	3.86	1.982	960.0	4.0	3.4	3.0	2.6	2.4	1.7	1.5	1.3	1.2		
33	3.74	2.119	990.0	3.7	3.2	2.8	2.5	2.2	1.6	1.4	1.2	1.1		
34	3.63	2.265	1020.0	3.5	3.0	2.6	2.3	2.1	1.5	1.3	1.1	1.0		
35	3.53	2.416	1050.0	3.3	2.8	2.5	2.2	2.0	1.4	1.2	1.1	1.0		
36	3.43	2.577	1080.0	3.1	2.7	2.3	2.1	1.9	1.3	1.1	1.0			
37	3.34	2.742	1110.0	3.0	2.5	2.2	2.0	1.8	1.2	1.1	1.0			
38	3.25	2.918	1140.0	2.8	2.4	2.1	1.9	1.7	1.2	1.0				
39	3.17	3.098	1170.0	2.7	2.3	2.0	1.7	1.6	1.1	1.0				
40	3.09	3.289	1200.0	2.5	2.2	1.9	1.6	1.5	1.1					

9″ "*Light*" *Beam.* Weight per lf. = 23.33 lbs.

Fig. 88.

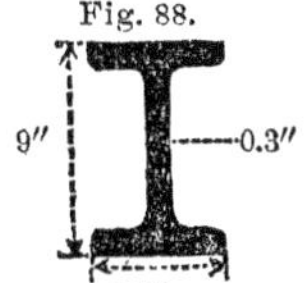

Sectional area......... = 7.0″
Moment of inertia I = 91.06
Constant K'............ = 101.2

$$W = \frac{K'}{L}.$$

Dis. bet. supports in feet.	Capac. W in tons.	Deflec. in inches.	Weight in lbs.	Distance d bet. centres of beams in feet, for weight in pounds per sq. foot of—										
				60 lbs.	70 lbs.	80 lbs.	90 lbs.	100 lbs.	140 lbs.	160 lbs.	180 lbs.	200 lbs.	250 lbs.	300 lbs.
6	16.86	0.062	140.0	......	......	......	......	......	......	......	......	......	22.0	18.7
7	14.45	0 085	163.3	......	......	......	......	......	......	25.0	22.0	20.6	16.6	13.7
8	12.65	0.111	186.6	......	......	......	......	......	25.0	19.7	17.5	15.8	12.6	10.5
9	11.24	0.141	210 0	......	......	......	......	......	17.8	15.6	13.8	12.5	10.1	8.6
10	10.12	0.175	233.3	......	......	......	22.0	20.0	14.4	12.6	11.2	10.1	8.0	6.7
11	9.20	0.212	256.6	......	......	21.0	18.7	16.7	11.9	10.4	9.2	8.3	6.7	5.5
12	8.43	0.253	280.0	23.0	20.0	17.5	15.6	14.0	10.0	8.7	7.8	7.0	5.6	4.6
13	7.78	0.297	303.3	19.9	17.9	14.9	13.4	11.9	8.5	7.4	6.6	5.9	4.8	3.9
14	7.22	0.345	326.6	17.1	14.7	12.8	11.4	10.3	7.3	6.4	5.7	5.1	4.1	3.4
15	6.74	0.398	350.0	14.9	12.9	11.2	10.0	8.9	6.4	5.6	5.0	4.5	3.6	2.9
16	6.31	0.453	373.3	13.1	11.2	9.8	8.7	7.8	5.6	4.9	4.3	3.9	3.1	2.6
17	5.95	0.510	396.6	11.6	10.0	8.7	7.8	7.0	5.0	4.3	3.8	3.5	2.8	2.3
18	5.62	0.579	420.0	10.4	8.9	7.8	6.9	6.2	4.4	3.9	3.4	3.1	2.5	2.0
19	5.32	0.648	443.3	9.3	8.0	7.0	6.2	5.6	4.0	3.5	3.1	2.8	2.2	1.8
20	5.06	0.721	466.6	8.4	7.2	6.3	5.6	5.0	3.6	3.1	2.8	2.5	2.0	1.6
21	4.81	0.797	490.0	7.6	6.5	5.7	5.1	4.5	3.2	2.8	2.5	2.2	1.8	1.5
22	4.59	0.879	513.3	6.9	5.9	5.2	4.6	4.1	2.9	2.6	2.3	2.1	1.7	1.3
23	4.40	0.968	536.6	6.3	5.5	4.7	4.2	3.8	2.7	2.3	2.1	1.9	1.5	1.2
24	4.21	1.060	560.0	5.8	5.0	4.3	3.8	3.5	2.5	2.1	1.9	1.7	1.4	1.1
25	4.04	1.151	583.3	5.3	4.6	4.0	3.5	3.2	2.3	2.0	1.7	1.6	1.2	1.0
26	3.89	1.254	606.6	4.9	4.2	3.7	3.3	2.9	2.1	1.8	1.6	1.4	1.1	
27	3.74	1.359	630.0	4.6	3.9	3.4	3.0	2.7	1.9	1.7	1.5	1.3	1 0	
28	3.60	1.466	653.3	4.2	3.6	3.2	2.8	2.5	1.8	1.6	1.4	1.2	1.0	
29	3.48	1.582	676.6	4.0	3.4	3.0	2.6	2.4	1.7	1.5	1.3	1.1		
30	3.37	1.711	700.0	3.7	3.2	2.8	2.5	2.2	1.6	1.4	1.2	1.1		
31	3.26	1.837	723.3	3.5	3.0	2.6	2.3	2.1	1.5	1.3	1.1	1.0		
32	3.16	1.968	746.6	3.2	2.8	2.4	2.1	1.9	1.4	1 2	1.0			
33	3.06	2.104	770.0	3.0	2.6	2.3	2.0	1.8	1.3	1.1				
34	2.97	2.250	793.3	2.9	2.4	2.1	1.9	1.7	1.2	1.0				
35	2.89	2.399	816.6	2.7	2.3	2.0	1.8	1.6	1.1					
36	2.81	2.565	840.0	2.6	2.2	1.9	1.7	1.5	1.0					
37	2.73	2.723	863.3	2.4	2.1	1.8	1.6	1.4						
38	2.66	2.898	886.6	2.3	2.0	1.7	1.5	1.4						
39	2.59	3.070	910.0	2.2	1.9	1.6	1.4	1.3						
40	2.52	3.250	933.3	2.1	1.8	1.5	1.4	1.2						

8″ *Beam.* Weight per lf. = 21.66 lbs.

Fig. 89.

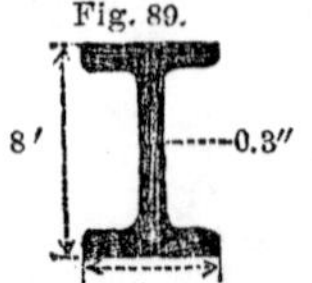

Sectional area......... = 6.5″
Moment of inertia I = 65.99
Constant K'............ = 82.49

$$W = \frac{K'}{L}.$$

Dis. bet. supports in feet.	Capac. W in tons.	Deflec. in inches.	Weight in lbs.	Distance d bet. centres of beams in feet, for weight in pounds per sq. foot of—										
				60 lbs.	70 lbs.	80 lbs.	90 lbs.	100 lbs.	140 lbs.	160 lbs.	180 lbs.	200 lbs.	250 lbs.	300 lbs.
6	13.74	0.070	130.0	......	......	......	......	......	......	......	......	22.0	18.3	15.2
7	11.78	0.095	151.6	......	......	......	......	......	......	21.0	18.6	16.8	13.4	11.2
8	10.30	0.124	173.3	......	......	......	......	25.7	18.3	16.0	14.3	12.8	10.3	8.5
9	9.16	0.158	195.0	......	......	......	......	20.3	14.5	12.7	11.3	10.1	8.1	6.7
10	8.23	0.198	216.6	......	......	20.5	18.3	16.4	11.7	10.2	9.1	8.2	6.5	5.4
11	7.49	0.238	238.3	22.6	19.4	17.0	15.1	13.6	9.7	8.5	7.5	6.8	5.4	4.5
12	6.87	0.284	260.0	19.0	16.3	14.3	12.7	11.4	8.1	7.1	6.3	5.7	4.5	3.8
13	6.34	0.335	281.6	16.2	14.0	12.1	10.9	9.7	6.9	6.0	5.4	4.8	3.9	3.2
14	5.89	0.390	303.3	14.0	12.0	10.5	9.3	8.4	6.0	5.2	4.6	4.2	3.3	2.8
15	5.49	0.447	325.0	12.2	10.5	9.1	8.1	7.3	5.2	4.5	4.0	3.6	2.9	2.4
16	5.15	0.511	346.6	10.7	9.1	8.0	7.1	6.4	4.5	4.0	3.6	3.2	2.5	2.1
17	4.85	0.580	368.3	9.5	8.0	7.1	6.3	5.7	4.0	3.7	3.2	2.8	2.3	1.9
18	4.58	0.653	390.0	8.4	7.2	6.3	5.6	5.1	3.8	3.1	2.7	2.5	2.0	1.7
19	4.34	0.731	411.6	7.6	6.5	5.7	5.1	4.5	3.3	2.9	2.5	2.2	1.8	1.5
20	4.11	0.810	433.3	6.8	5.8	5.1	4.5	4.1	2.9	2.5	2.2	2.0	1.6	1.3
21	3.92	0.898	455.0	6.2	5.3	4.6	4.1	3.7	2.6	2.3	2.0	1.8	1.4	1.2
22	3.73	0.989	476.6	5.6	4.8	4.2	3.7	3.4	2.4	2.1	1.8	1.6	1.3	1.1
23	3.58	1.090	498.3	5.1	4.4	3.8	3.4	3.1	2.2	1.9	1.7	1.5	1.2	1.0
24	3.42	1.192	520.0	4.7	4.0	3.5	3.1	2.8	2.0	1.7	1.5	1.4	1.1	
25	3.29	1.300	541.6	4.3	3.7	3.2	2.9	2.6	1.8	1.5	1.4	1.3	1.0	
26	3.17	1.417	563.3	4.0	3.4	3.0	2.7	2.4	1.7	1.4	1.3	1.2		
27	3.05	1.536	585.0	3.7	3.2	2.8	2.5	2.2	1.6	1.4	1.2	1.1		
28	2.94	1.662	606.6	3.5	3.0	2.6	2.3	2.1	1.5	1.3	1.1	1.0		
29	2.84	1.795	628.3	3.2	2.8	2.4	2.1	1.9	1.4	1.2	1.0			
30	2.73	1.923	650.0	3.0	2.6	2.2	2.0	1.8	1.3	1.1				
31	2.66	2.080	671.6	2.8	2.4	2.1	1.9	1.7	1.2	1.0				
32	2.56	2.219	693.3	2.6	2.2	2.0	1.7	1.6	1.1					
33	2.49	2.383	715.0	2.5	2.1	1.8	1.6	1.4	1.0					
34	2.42	2.550	736.6	2.3	2.0	1.7	1.5	1.4						
35	2.35	2.722	758.3	2.2	1.9	1.6	1.4	1.3						
36	2.29	2.907	780.0	2.1	1.8	1.5	1.4	1.2						
37	2.22	3.084	801.6	2.0	1.7	1.5	1.3	1.2						
38	2.17	3.290	823.3	1.9	1.6	1.4	1.2	1.1						
39	2.11	3.484	845.0	1.8	1.5	1.3	1.2	1.0						
40	2.06	3.702	866.6	1.7	1.4	1.2	1.1	1.0						

Fig. 90.

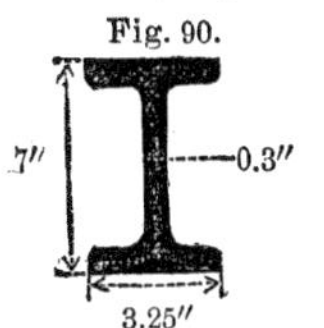

7″ Beam. Weight per lf. = 18.33 lbs.

Sectional area......... = 5.5″
Moment of inertia I = 44.84
Constant K'........... = 64.06

$$W = \frac{K'}{L}.$$

Dis. bet. supports in feet.	Capac. W in tons.	Deflec. in inches.	Weight in lbs.	Distance d bet. centres of beams in feet, for weight in lbs. per sq. foot of—										
				60 lbs.	70 lbs.	80 lbs.	90 lbs.	100 lbs.	140 lbs.	160 lbs.	180 lbs.	200 lbs.	250 lbs.	300 lbs.
6	10.67	0.080	110.0	......	......	......	.....	......	25.4	22 2	19.7	17.7	14.2	11.8
7	9.15	0.109	128.3	......	......	......	......	......	18.6	16.3	14.5	13.0	10.5	8.7
8	8.00	0.143	146.6	......	......	25.0	22.2	20.0	14.2	12.5	11.1	10.0	8.0	6.6
9	7.11	0.181	165.0	......	22.9	19.7	17.5	15.8	11.2	9.8	8.7	7.9	6.3	5.2
10	6.40	0.224	183.3	21.3	18.2	16.0	14.2	12.8	9.1	8.0	7.1	6.4	5.1	4.2
11	5.82	0.272	201.6	17.6	15.3	13.2	11.7	10.5	7.5	6.6	5.8	5.2	4.2	3.5
12	5.33	0.325	220.0	14.8	12.6	11.1	9.8	8.8	6.3	5.5	4.9	4.4	3.5	2.9
13	4.92	0.382	238.3	12.6	10.9	9.4	8.8	7.5	5.4	4.7	4.1	3.7	3.0	2.5
14	4.56	0.444	256.6	10.8	9.3	8.1	7.2	6.5	4.6	4.0	3.6	3.2	2.6	2.1
15	4.27	0.513	275.0	9.4	8.2	7.1	6.3	5.7	4.0	3 5	3.1	2.8	2.2	1.8
16	3.99	0.585	293.3	8.3	7.1	6.2	5.5	4.9	3.5	3.1	2.7	2.4	1.9	1.6
17	3.76	0.665	311.6	7.3	6.5	5.5	4.9	4.4	3.1	2.7	2.3	2.1	1.7	1.4
18	3.55	0.749	330.0	6.5	5.6	4.9	4.3	3.9	2.8	2.4	2.1	1.9	1.5	1.3
19	3.37	0.840	348.3	5.9	5.1	4.4	3.9	3.5	2.5	2.2	1.9	1.7	1.4	1.1
20	3.20	0.936	366.6	5.3	4.5	4.0	3.5	3.2	2.2	2.0	1.7	1.6	1.2	1.0
21	3.05	1.038	385.0	4.8	4.1	3.6	3.2	2.9	2.0	1.8	1.6	1.4	1.1	
22	2.91	1.146	403.3	4.4	3.7	3.3	2.9	2.6	1.8	1.6	1.4	1.3	1.0	
23	2.78	1.257	421.6	4.0	3.4	3.0	2.7	2.4	1.7	1.5	1.3	1.2		
24	2.66	1.381	440.0	3.6	3.1	2.7	2.4	2.2	1.6	1.3	1.2	1.1		
25	2.56	1.504	458.3	3.4	2.9	2.5	2.2	2.0	1.5	1.2	1.1	1.0		
26	2.45	1.630	476.6	3.1	2.6	2.3	2.0	1.8	1.4	1.1				
27	2.37	1.775	495.0	2.9	2.5	2.1	1.9	1.7	1.3	1.0				
28	2.27	1.871	513.3	2.7	2.3	2.0	1.8	1.6	1.2					
29	2.20	2.075	531.6	2.5	2.1	1.8	1.7	1.5	1.1					
30	2.12	2.229	550.0	2.3	2.0	1.7	1.5	1.4	1.0					

Fig. 91.

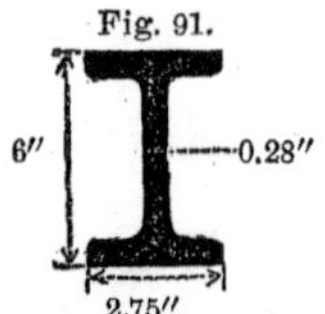

6″ *Beam*. Weight per lf. = 13.33 lbs.

Sectional area......... = 4.0″
Moment of inertia I = 22.5
Constant K'........... = 37.64

$$W = \frac{K'}{L}.$$

Dis. bet. supports in feet.	Capac. W in tons.	Deflec. in inches.	Weight in lbs.	Distance d bet. centres of beams in feet, for weight in lbs. per sq. foot of—										
				60 lbs.	70 lbs.	80 lbs.	90 lbs.	100 lbs.	140 lbs.	160 lbs.	180 lbs.	200 lbs.	250 lbs.	300 lbs.
6	6.27	0.094	80.0	......	......	......	23.2	20.9	14.9	13.0	11.6	10.4	8.3	6.9
7	5.37	0.128	93.3	......	......	19.1	17.3	15.3	10.0	9.5	8.4	7.6	6.1	5.1
8	4.70	0.168	106.6	19.5	16.8	14.6	13.0	11.7	8.5	7.3	6.5	5.8	4.7	3.9
9	4.18	0.213	120.0	15.4	13.4	11.6	10.4	9.2	6.6	5.8	5.1	4.6	3.7	3.1
10	3.75	0.263	133.3	12.5	10.7	9.3	8.3	7.5	5.3	4.7	4.1	3.7	3.0	2.5
11	3.42	0.320	146.6	10.3	9.0	7.7	6.9	6.2	4.4	3.8	3.4	3.1	2.4	2.0
12	3.13	0.382	160.0	8.6	7.0	6.5	5.7	5.2	3.7	3.2	2.9	2.6	2.0	1.7
13	2.89	0.450	173.3	7.4	6.4	5.5	4.9	4.4	3.1	2.7	2.4	2.2	1.7	1.4
14	2.68	0.524	186.6	6.3	5.4	4.7	4.2	3.8	2.7	2.3	2.1	1.9	1.5	1.2
15	2.51	0.607	200.0	5.5	4.8	4.2	3.7	3.3	2.3	2.1	1.8	1.6	1.3	1.1
16	2.34	0.689	213.3	4.8	4.1	3.6	3.2	2.9	2.0	1.8	1.6	1.4	1.1	
17	2.21	0.786	226.6	4.3	3.7	3.2	2.9	2.5	1.8	1.6	1.4	1.3		
18	2.09	0.888	240.0	3.8	3.3	2.9	2.5	2.3	1.6	1.4	1.2	1.1		
19	1.98	0.995	253.3	3.4	3.0	2.6	2.3	2.1	1.4	1.3	1.1			
20	1.88	1.110	266.6	3.1	2.7	2.3	2.1	1.8	1.3	1.1				
21	1.79	1.231	280.0	2.8	2.4	2.1	1.9	1.7	1.2	1.0				
22	1.70	1.350	293.3	2.5	2.2	1.9	1.7	1.5	1.1					
23	1.63	1.493	306.6	2.3	2.0	1.7	1.5	1.4	1.0					
24	1.56	1.641	320.0	2.1	1.8	1.6	1.4	1.3						
25	1.50	1.787	333.3	2.0	1.7	1.5	1.3	1.2						
26	1.44	1.950	346.6	1.8	1.5	1.3	1.2	1.1						
27	1.39	2.129	360.0	1.7	1.4	1.2	1.1							
28	1.33	2.286	373.3	1.5	1.3	1.1								
29	1.29	2.489	386.6	1.4	1.2	1.0								
30	1.25	2.698	400.0	1.3	1.1									

CAST-IRON BEAMS.

Factor of rupture C for cast-iron beams of various sections.

The factor C is based on practical experiments by Hodgkinson. Its value alters with the different proportions of the cross-sections of beam.

Beam supported at the ends; load concentrated at the center.

Reference.

C = Factor of rupture.
W = Breaking weight in lbs.
A = Sectional area of beam in square inches.
l = Distance between supports in inches.
h = Height of beam in inches.

$$C = \frac{W.l}{A.h}, \quad W = \frac{A.h}{l}.C.$$

Dimensions in inches. b = Thickness of web at center is the unit.

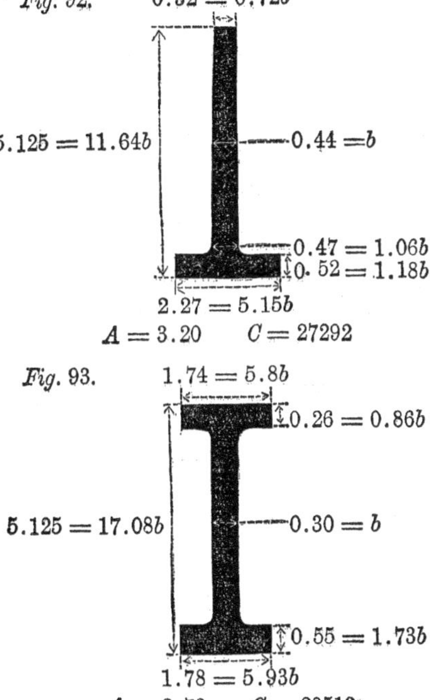

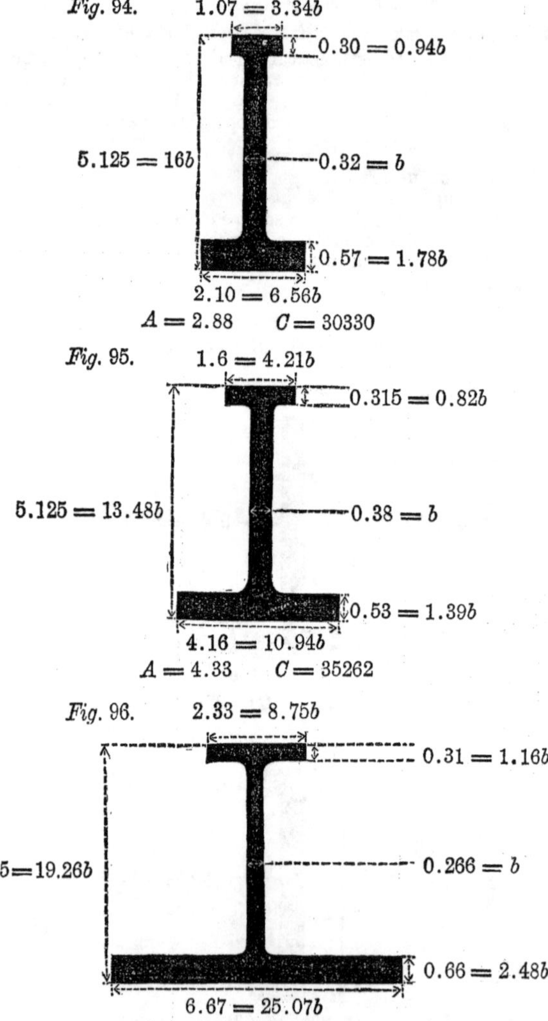
Fig. 94.
1.07 = 3.34b
0.30 = 0.94b
5.125 = 16b
0.32 = b
0.57 = 1.78b
2.10 = 6.56b
A = 2.88 C = 30330
Fig. 95.
1.6 = 4.21b
0.315 = 0.82b
5.125 = 13.48b
0.38 = b
0.53 = 1.39b
4.16 = 10.94b
A = 4.33 C = 35262
Fig. 96.
2.33 = 8.75b
0.31 = 1.16b
5.125=19.26b
0.266 = b
0.66 = 2.48b
6.67 = 25.07b
A = 6.23 C = 44176

Theoretical cross-section of equal resistance, according to Moll and Reuleaux.

The following are theoretically the most economical sections. They form the basis for the table "Capacity of Cast-iron Beams." Flange nearest the neutral axis mn in tension:

No. of Figure.	Form of section.	Moment of inertia I.	Moment of resistance $\frac{I}{s}$	Sectional area A in inches.
97	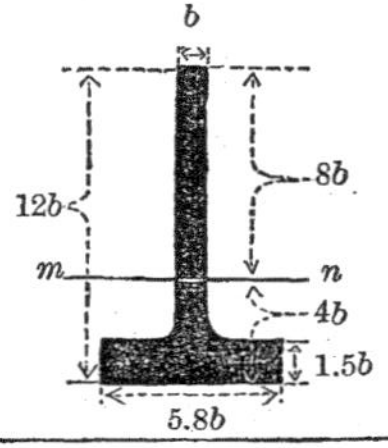	$269.4b^4$	$33.675b^3$	$19.2b^2$
98	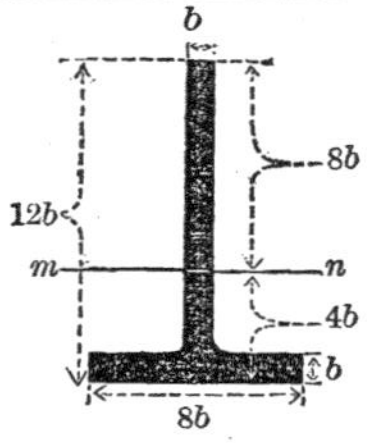	$278b^4$	$34.8b^3$	$19b^2$

Theoretical cross-section of equal resistance—Continued.

No. of Figure.	Form of section.	Moment of inertia I.	Moment of resistance $\frac{I}{s}$	Sectional area A in inches.
99	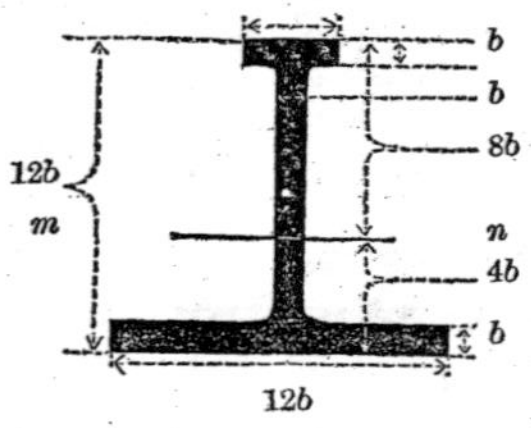	$440b^4$	$55b^3$	$25b^2$
100	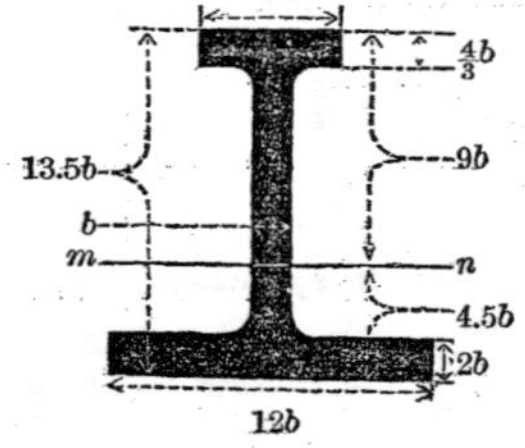	$922b^4$	$102.4b^3$	$40.82b^2$

101	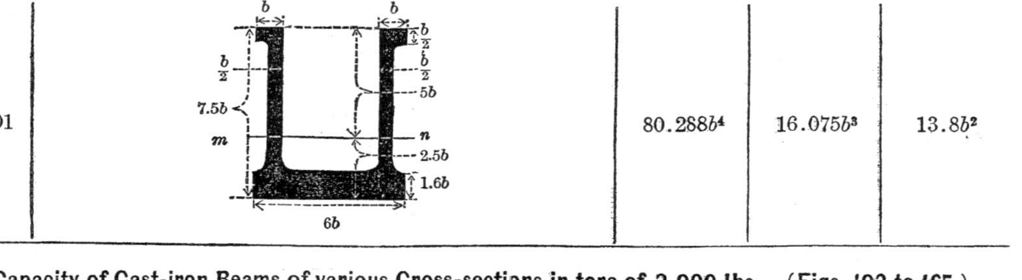	$80.288b^4$	$16.075b^3$	$13.8b^2$

Capacity of Cast-iron Beams of various Cross-sections in tons of 2,000 lbs. (Figs. 102 to 165.)

The following table gives the moment of resistance multiplied by the modulus of rupture:

The capacity W is equal to the tabulated coefficient K^1 divided by the distance l between supports, as shown in formulas below for variously circumstanced beams. For beams fixed at one end, l is distance between support and W at free end of beam.

The modulus of rupture is taken at $\frac{28000}{4} = 7{,}000$ lbs. at 4 times safety.

All dimensions in inches.

Capacity W in tons of 2,000 lbs.

Beam supported at both ends; principal flange at bottom.

Load equally distributed: $W = \frac{K^1}{\frac{1}{2}.l}$, or $K^1 = \frac{l.W}{2}$.

Load concentrated at center: $W = \frac{K^1}{l}$, or $K^1 = l.W.$

Beam fixed at one end; principal flange at top.

Load equally distributed: $W = \frac{K^1}{2.l}$, or $K^1 = 2.l.W.$

Load concentrated at free end: $W = \frac{K^1}{4.l}$, or $K^1 = 4.l.W.$

[NOTE.—The more the sectional area is contained in coefficient K^1, the more is the section economical.]

EXAMPLE.—Section No. 34. Load equally distributed; beam supported at both ends; thickness of web = 1 inch; thickness of flange = $1\frac{1}{4}$ inch; height = 10 inches; width of flange = 5.9 inches. Distance between supports = 20 feet = 240 inches.

$$W = \frac{K^1}{\frac{1}{2}l} = \frac{658}{120} = 5.48 \text{ tons capacity.}$$

The moment of resistance of cross-section $\frac{I}{s} = \frac{K^1}{14}$

Fig. 102.

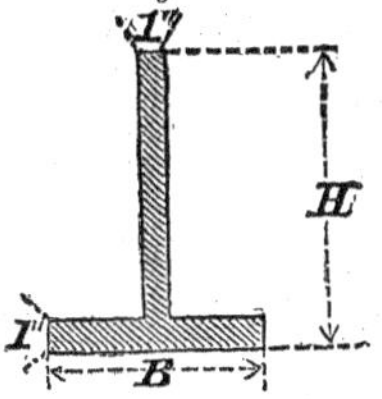

Fig. 103.

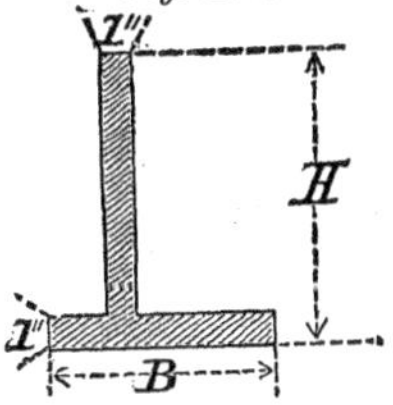

Fig. 104.

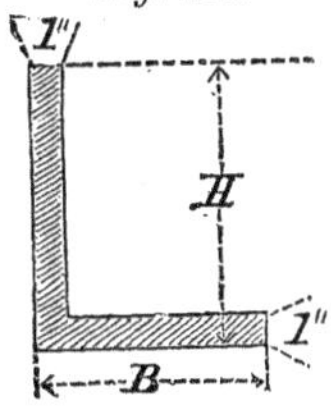

Fig. 105.

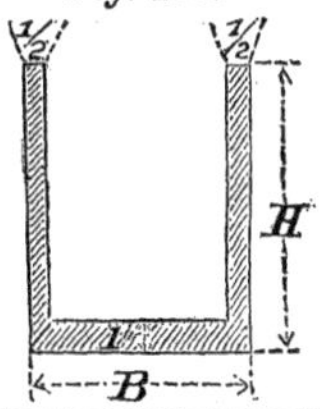

Number of section.	Height H in inches.	Width B of lower flange in inches.	Sectional area in square inches.	Coefficient K^1
1	6	5.0	10.0	238
2	6½	5.2	10.7	280
3	7	5.5	11.5	322
4	7½	5.7	12.2	364
5	8	6.0	13.0	420
6	8½	6.2	13.7	476
7	9	6.5	14.5	532
8	9½	6.7	15.2	602
9	10	6.9	15.9	672
10	10½	7.1	16.6	742
11	11	7.4	17.4	812
12	11½	7.6	18.1	882
13	12	7.9	18.9	966
14	12½	8.1	19.6	1050
15	13	8.4	20.4	1134
16	13½	8.6	21.1	1232
17	14	8.8	21.8	1316
18	14½	9.0	22.5	1428
19	15	9.3	23.3	1526
20	15½	9.5	24.0	1624
21	16	9.8	24.8	1750
22	16½	10.0	25.5	1848
23	17	10.3	26.3	1960
24	17½	10.5	27.0	2086
25	18	10.8	27.8	2212

Fig. 106.

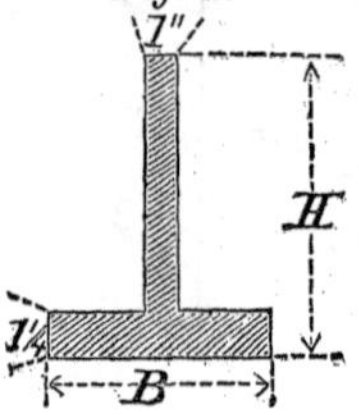

Fig. 107.

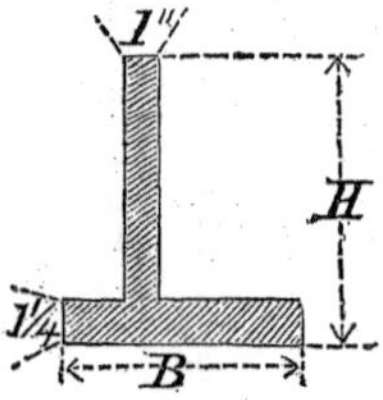

Fig. 108.

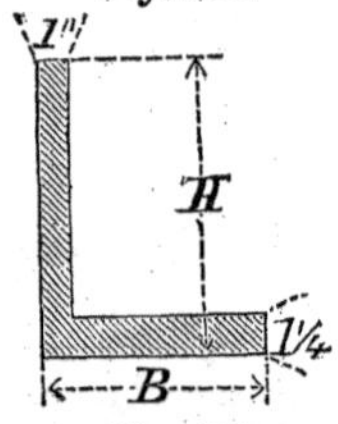

Fig. 109.

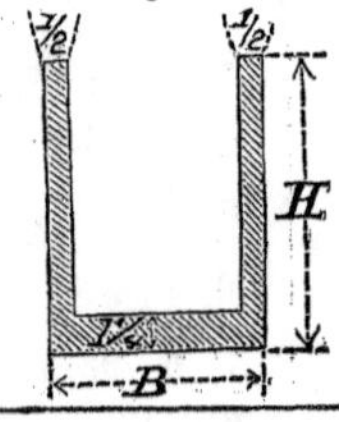

Number of section.	Height H in inches.	Width B of lower flange in inches.	Sectional area in square inches.	Coefficient $K1$
26	6	4.5	10.4	224
27	6½	4.6	11.1	266
28	7	4.8	11.8	322
29	7½	5.0	12.5	364
30	8	5.2	13.2	420
31	8½	5.4	13.9	476
32	9	5.6	14.7	532
33	9½	5.7	15.4	588
34	10	5.9	16.2	658
35	10½	6.1	16.9	728
36	11	6.3	17.6	798
37	11½	6.5	18.3	882
38	12	6.7	19.1	952
39	12½	6.9	19.8	1036
40	13	7.1	20.6	1134
41	13½	7.3	21.3	1218
42	14	7.5	22.1	1316
43	14½	7.7	22.8	1414
44	15	7.9	23.6	1512
45	15½	8.0	24.3	1610
46	16	8.2	25.1	1722
47	16½	8.4	25.8	1834
48	17	8.6	26.5	1946
49	17½	8.8	27.2	2072
50	18	9.0	28.0	2198

Fig. 110.

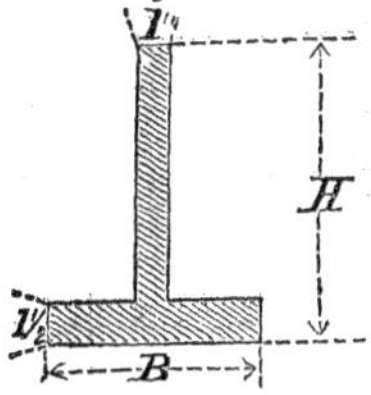

Fig. 111.

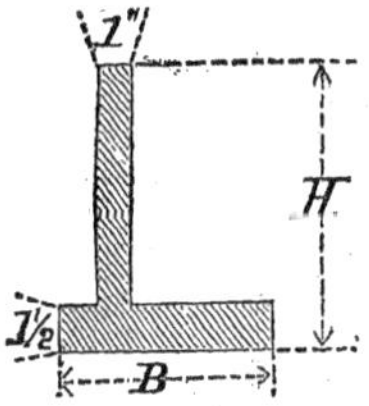

Fig. 112.

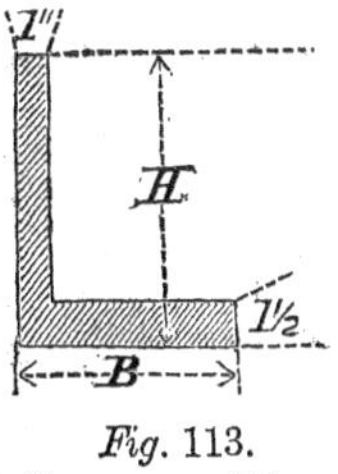

Fig. 113.

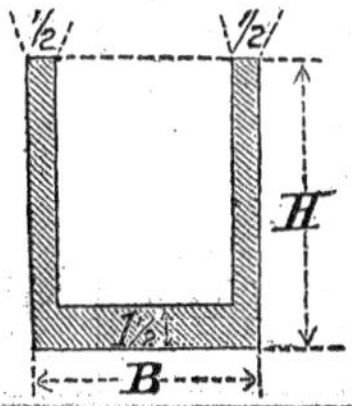

Number of section.	Height H in inches.	Width B of lower flange in inches.	Sectional area in square inches.	Coefficient K_1
51	6	4.2	10.5	224
52	6½	4.3	11.4	266
53	7	4.5	12.3	308
54	7½	4.6	12.9	364
55	8	4.7	13.6	406
56	8½	4.8	14.3	462
57	9	5.0	15.0	532
58	9½	5.1	15.7	588
59	10	5.3	16.5	658
60	10½	5.4	17.2	728
61	11	5.6	17.9	798
62	11½	5.7	18.6	868
63	12	5.9	19.4	952
64	12½	6.0	20.1	1036
65	13	6.3	20.9	1120
66	13½	6.4	21.6	1204
67	14	6.6	22.4	1302
68	14½	6.7	23.1	1400
69	15	6.9	23.8	1498
70	15½	7.0	24.5	1610
71	16	7.2	25.3	1708
72	16½	7.3	26.0	1820
73	17	7.5	26.8	1932
74	17½	7.7	27.5	2058
75	18	7.9	28.3	2184

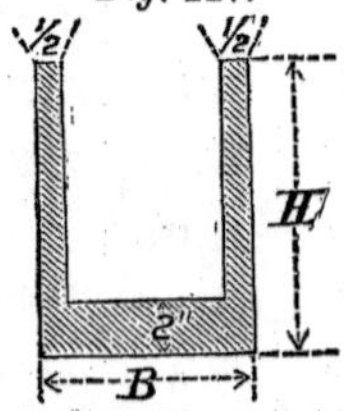

Number of section.	Height H in inches.	Width B of lower flange in inches.	Sectional area in square inches.	Coefficient K.1
76	6	4.0	12.0	224
77	7	4.1	13.1	308
78	8	4.2	14.4	406
79	9	4.4	15.7	518
80	10	4.6	17.1	644
81	11	4.8	18.6	784
82	12	5.0	20.0	938
83	13	5.2	21.4	1106
84	14	5.5	22.9	1288
85	15	5.7	24.4	1484
86	16	5.9	25.8	1694
87	17	6.2	27.3	1918
88	18	6.4	28.8	2156

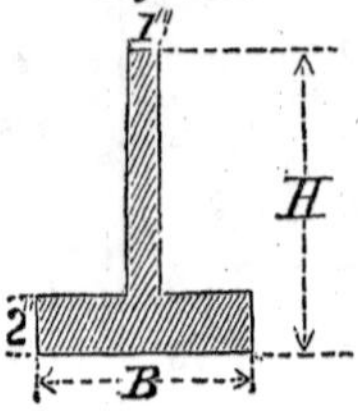

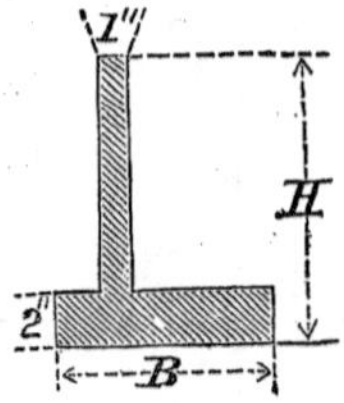

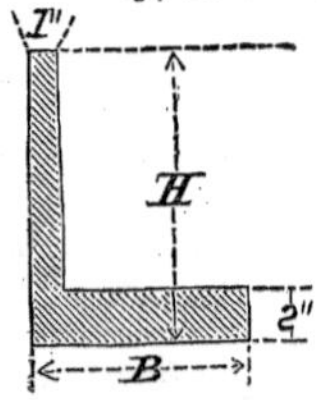

Fig. 114.

Fig. 115.

Fig. 116.

Fig. 117.

Fig. 118.

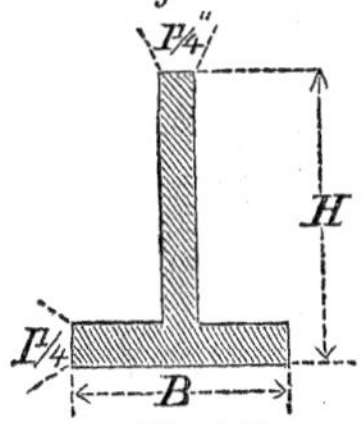

Fig. 119.

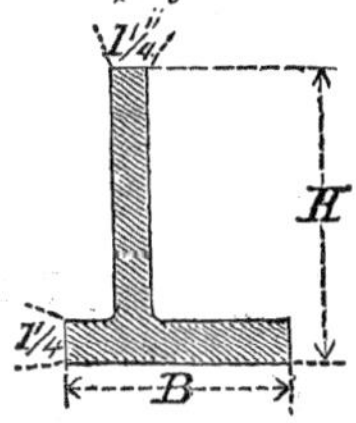

Fig. 120.

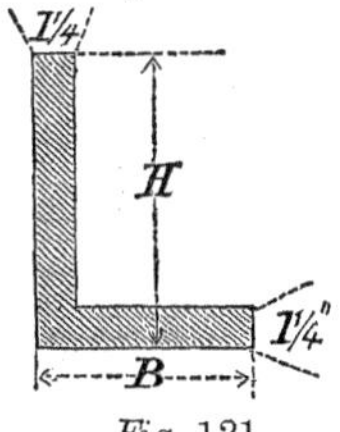

Fig. 121.

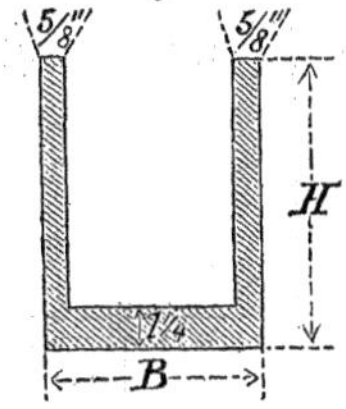

Number of section.	Height *H* in inches.	Width *B* of lower flange in inches.	Sectional area in square inches.	Coefficient *K*1.
89	6	5.6	12.9	294
90	6½	5.8	13.8	336
91	7	6.0	14.7	392
92	7½	6.2	15.5	448
93	8	6.4	16.4	518
94	8½	6.6	17.3	588
95	9	6.9	18.3	658
96	9½	7.1	19.2	742
97	10	7.4	20.2	826
98	10½	7.6	21.1	910
99	11	7.9	22.1	1008
100	11½	8.1	23.0	1106
101	12	8.4	23.9	1204
102	12½	8.6	24.8	1302
103	13	8.9	25.8	1414
104	13½	9.1	26.7	1526
105	14	9.4	27.7	1652
106	14½	9.6	28.5	1764
107	15	9.8	29.4	1890
108	15½	10.0	30.3	2030
109	16	10.3	31.3	2156
110	16½	10.5	32.2	2296
111	17	10.8	33.2	2436
112	17½	11.0	34.1	2590
113	18	11.3	35.0	2730

Fig. 122.

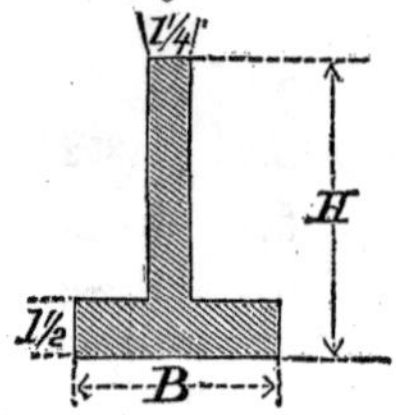

Fig. 123.

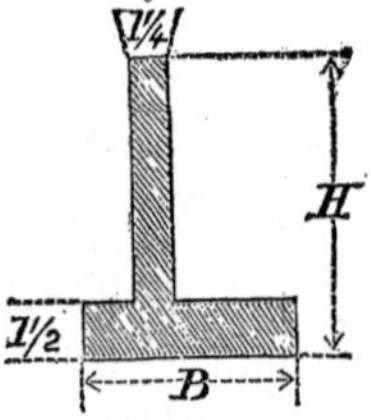

Fig. 124.

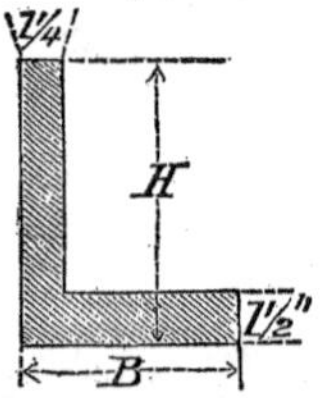

Fig. 125.

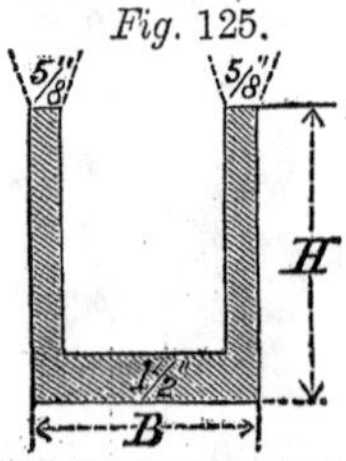

Number of section.	Height H in inches.	Width B of lower flange in inches.	Sectional area in square inches.	Coefficient $K1$.
114	6	5.3	13.6	280
115	6½	5.4	14.4	336
116	7	5.6	15.3	392
117	7½	5.7	16.1	448
118	8	5.9	17.0	518
119	8½	6.0	17.8	588
120	9	6.2	18.7	658
121	9½	6.4	19.6	742
122	10	6.6	20.5	814
123	10½	6.8	21.4	910
124	11	7.0	22.4	994
125	11½	7.2	23.3	1092
126	12	7.4	24.2	1190
127	12½	7.6	25.1	1288
128	13	7.8	26.1	1400
129	13½	8.0	27.0	1512
130	14	8.2	27.9	1624
131	14½	8.4	28.8	1750
132	15	8.6	29.8	1876
133	15½	8.8	30.7	2002
134	16	9.0	31.6	2142
135	16½	9.2	32.5	2282
136	17	9.4	33.5	2422
137	17½	9.6	34.4	2562
138	18	9.8	35.3	2716

Fig. 126.

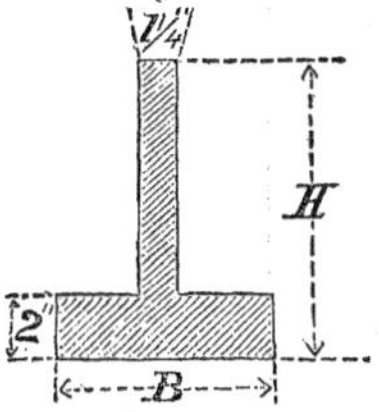

Fig. 127.

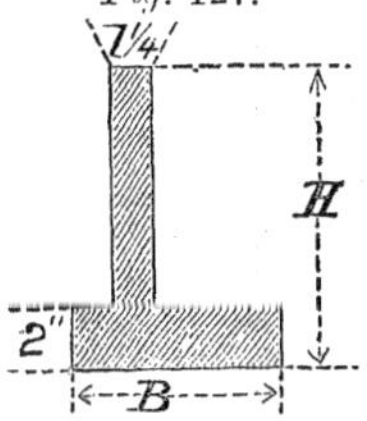

Fig 128.

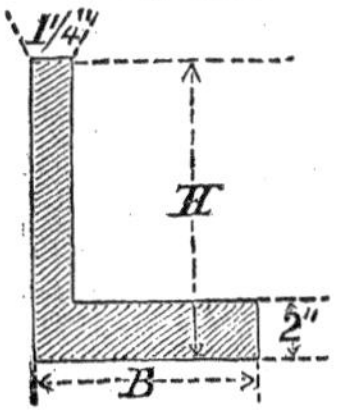

Fig. 129.

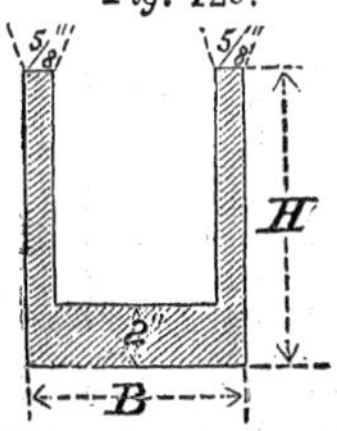

Number of section.	Height H in inches.	Width B of lower flange in inches.	Sectional area in square inches.	Coefficient K^1.
139	6	5.0	15.0	280
140	7	5.1	16.4	378
141	8	5.3	18.0	504
142	9	5.5	19 7	644
143	10	5.7	21.4	798
144	11	6.0	23.2	980
145	12	6.3	25.0	1162
146	13	6.5	26.8	1372
147	14	6.8	28.6	1610
148	15	7.1	30.5	1848
149	16	7.4	32.3	2114
150	17	7.7	34.2	2394
151	18	8.0	36.0	2688

Fig. 130.

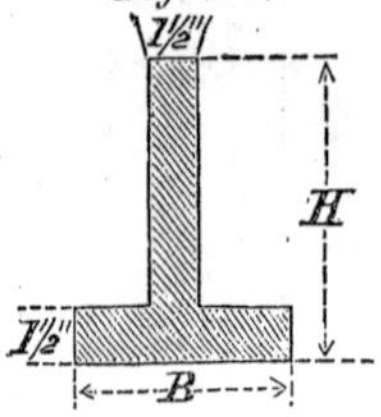

Fig. 131.

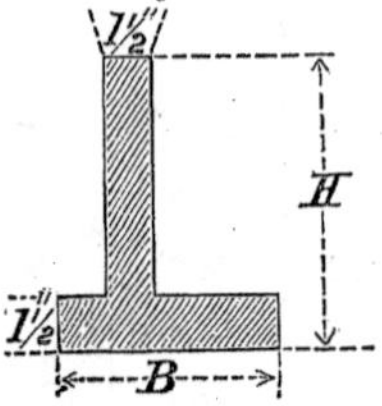

Fig. 132.

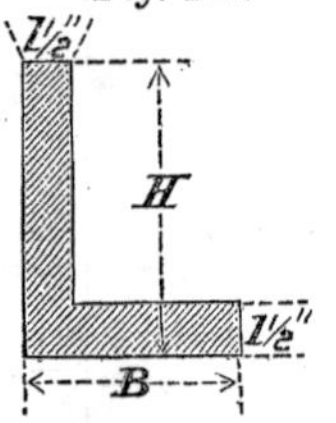

Fig. 133.

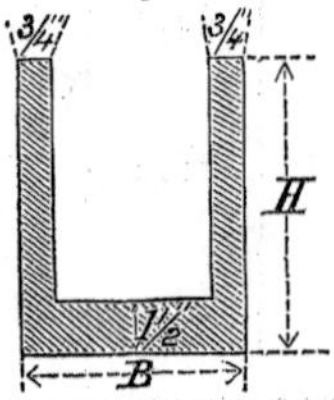

Number of section.	Height H in inches.	Width B of lower flange in inches.	Sectional area in square inches.	Coefficient K^1.
152	6	6.3	16.2	336
153	6½	6.5	17.2	406
154	7	6.7	18.3	476
155	7½	6.9	19.3	546
156	8	7.1	20.4	616
157	8½	7.3	21.5	700
158	9	7.5	22.6	784
159	9½	7.7	23.6	882
160	10	8.0	24.7	980
161	10½	8.2	25.8	1078
162	11	8.4	26.9	1190
163	11½	8.6	28.0	1302
164	12	8.9	29.1	1428
165	12½	9.1	30.1	1554
166	13	9.3	31.2	1680
167	13½	9.5	32.3	1806
168	14	9.8	33.5	1960
169	14½	10.0	34.6	2100
170	15	10.3	35.7	2254
171	15½	10.5	36.8	2408
172	16	10.8	38.0	2562
173	16½	11.0	39.1	2730
174	17	11.3	40.2	2912
175	17½	11.5	41.3	3080
176	18	11.8	42.5	3262

Fig. 134.

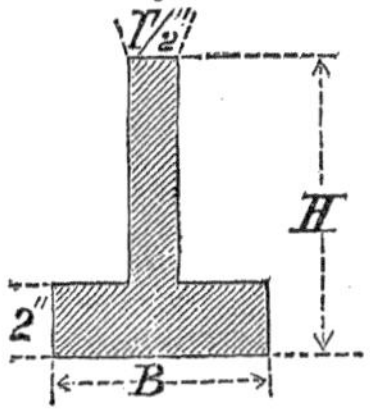

Fig. 135.

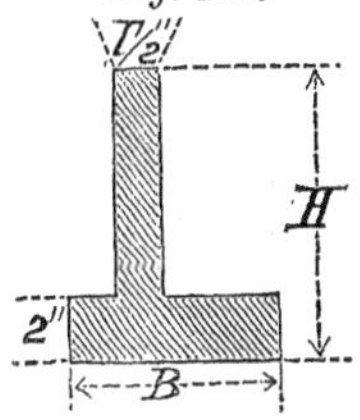

Fig. 136.

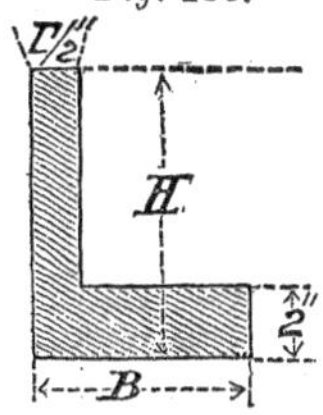

Fig. 137.

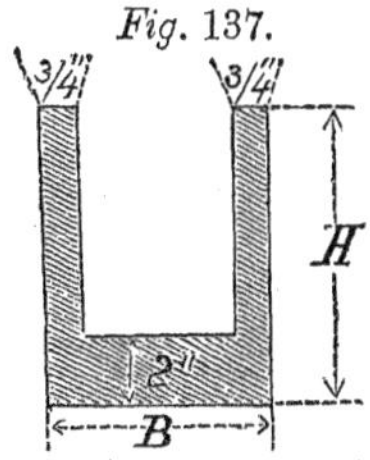

Number of section.	Height H in inches.	Width B of lower flange in inches.	Sectional area in square inches.	Coefficient K^1.
177	6	6.0	18.0	336
178	7	6.1	19.7	462
179	8	6.3	21.6	602
180	9	6.6	23.6	770
181	10	6.9	25.7	966
182	11	7.2	27.9	1176
183	12	7.5	30.0	1400
184	13	7.8	32.2	1652
185	14	8.2	34.4	1932
186	15	8.5	36.7	2212
187	16	8.9	38.8	2534
188	17	9.2	41.0	2370
189	18	9.6	43.2	3220

Fig. 138.

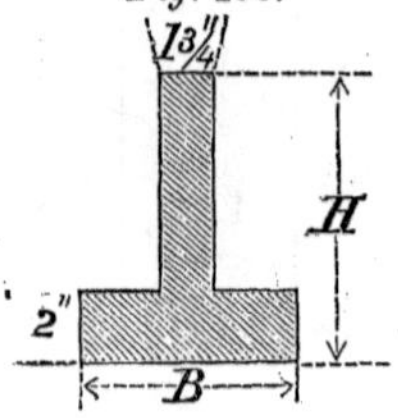

Fig. 139.

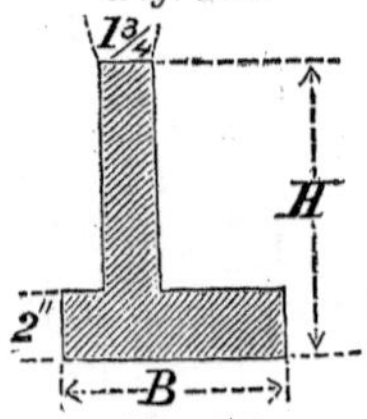

Fig. 140.

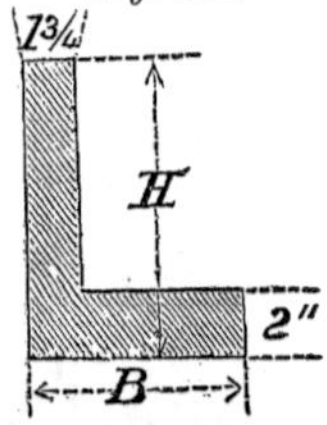

Fig. 141.

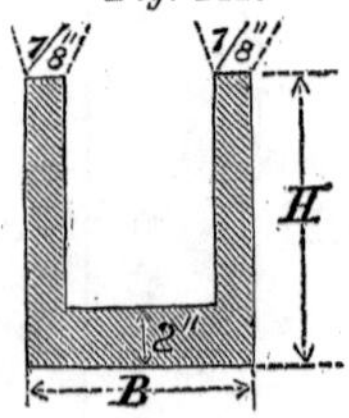

Number of section.	Height H in inches.	Width B of lower flange in inches.	Sectional area in square inches.	Coefficient K^1.
190	6	7.0	21.0	392
191	7	7.1	23.0	532
192	8	7.4	25.2	714
193	9	7.7	27.6	896
194	10	8.0	30.0	1120
195	11	8.4	32.5	1372
196	12	8.8	35.0	1638
197	13	9.1	37.5	1932
198	14	9.6	40.1	2240
199	15	10.0	42.7	2590
200	16	10.4	45.2	2954
201	17	10.8	47.8	3346
202	18	11.2	50.4	3766

Fig. 142.

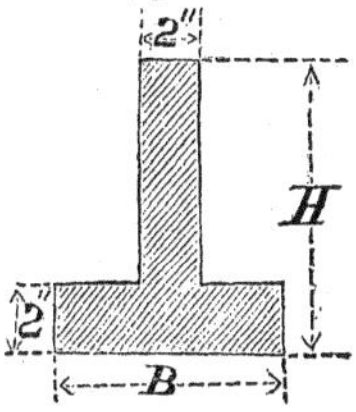

Fig. 143.

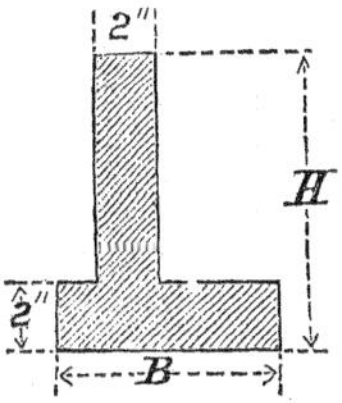

Fig. 144.

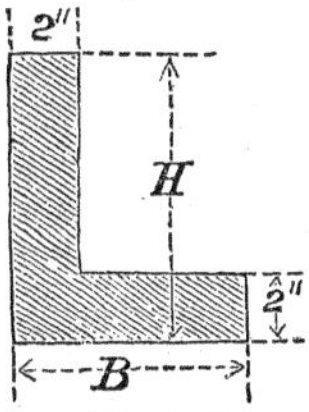

Fig. 145.

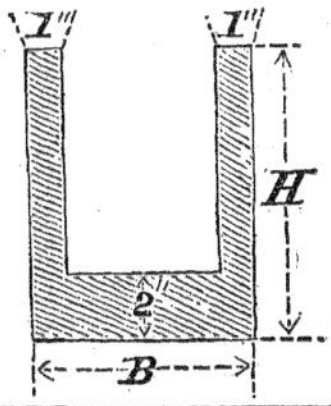

Number of section.	Height H in inches.	Width B of lower flange in inches.	Sectional area in square inches.	Coefficient $K1$.
203	6	8.0	24.0	448
204	7	8.1	26.2	616
205	8	8.4	28.8	812
206	9	8.8	31.5	1036
207	10	9.1	34.3	1274
208	11	9.6	37.1	1554
209	12	10.0	40.0	1862
210	13	10.4	42.9	2198
211	14	10.9	45.8	2562
212	15	11.4	48.7	2954
213	16	11.8	51.7	3374
214	17	12.3	54.6	3822
215	18	12.8	57.6	4298

Fig. 146.

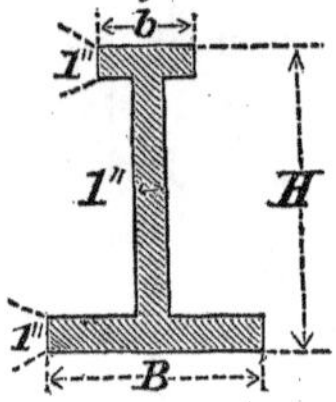

Fig. 147.

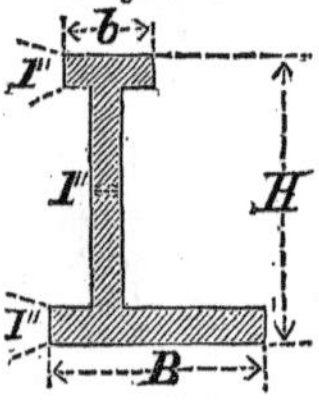

Fig. 148.

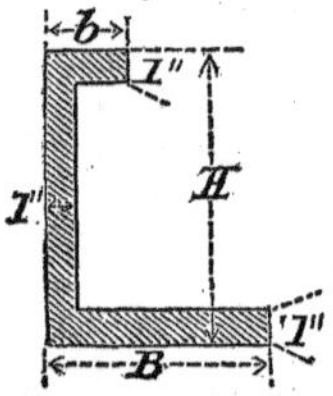

Fig. 149.

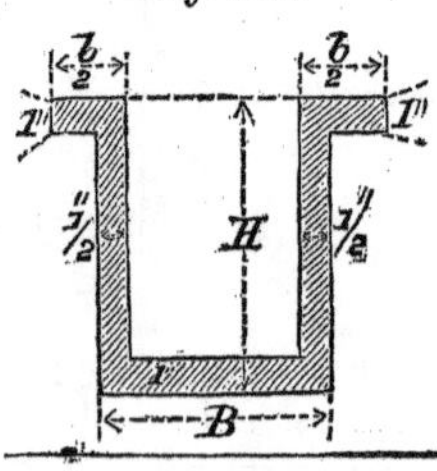

Number of section.	Height H in inches.	Width B of lower flange in inches.	Width b of upper flange in inches.	Sectional area in square inches.	Coefficient K^1.
1	6	6	1.4	11.4	294
2	6	7	1.9	12.9	336
3	6	8	2.3	14.3	392
4	6	9	2.7	15.7	448
5	6	10	3.1	17.1	504
6	6	11	3.6	18.6	560
7	6	12	4.0	20.0	602
8	6	13	4.4	21.4	658
9	6	14	4.9	22.9	714
10	6	15	5.3	24.3	770
11	6	16	5.7	25.7	826
12	6	17	6.2	27.2	868
13	6	18	6.6	28.6	924
14	7	6	1.2	12.2	350
15	7	7	1.7	13.7	420
16	7	8	2.1	15.1	490
17	7	9	2.6	16.6	560
18	7	10	3.0	18.0	616
19	7	11	3.4	19.4	686
20	7	12	3.9	20.9	756
21	7	13	4.3	22.3	826
22	7	14	4.8	23.8	896
23	7	15	5.2	25.2	966
24	7	16	5.7	26.7	1022
25	7	17	6.1	28.1	1092

Fig. 146.

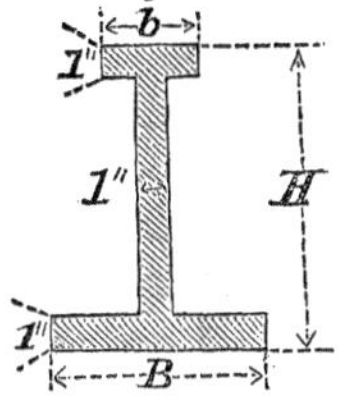

Fig. 147.

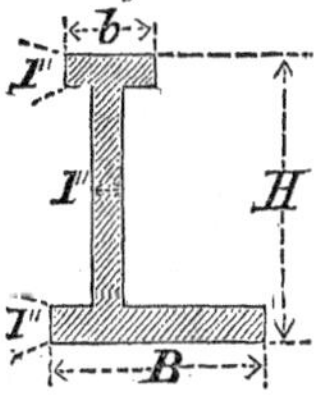

Fig. 148.

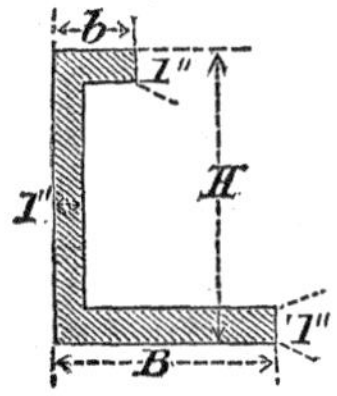

Fig. 149.

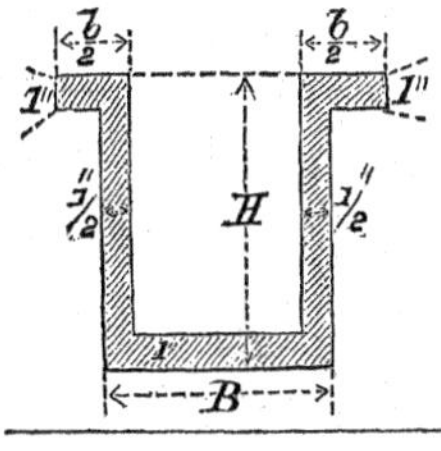

Number of section.	Height H in inches.	Width B of lower flange in inches.	Width b of upper flange in inches.	Sectional area in square inches.	Coefficient K^1.
26	7	18	6.5	29.5	1162
27	8	6	1.0	13.0	434
28	8	7	1.5	14.5	504
29	8	8	1.9	15.9	588
30	8	9	2.4	17.4	672
31	8	10	2.8	18.8	742
32	8	11	3.3	20.3	826
33	8	12	3.7	21.7	910
34	8	13	4.2	23.2	994
35	8	14	4.6	24.6	1078
36	8	15	5.1	26.1	1148
37	8	16	5.5	27.5	1232
38	8	17	6.0	29.0	1316
39	8	18	6.4	30.4	1386
40	9	7	1.3	15.3	588
41	9	8	1.7	16.7	686
42	9	9	2.2	18.2	784
43	9	10	2.6	19.6	868
44	9	11	3.1	21.1	966
45	9	12	3.5	22.5	1064
46	9	13	4.1	24.1	1162
47	9	14	4.5	25.5	1246
48	9	15	4.9	26.9	1344
49	9	16	5.4	28.4	1442
50	9	17	5.8	29.8	1526

Fig. 146.

Fig. 147.

Fig. 148.

Fig. 149.

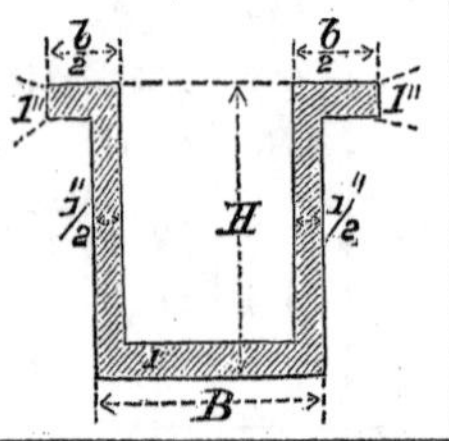

Number of section.	Height H in inches.	Width B of lower flange in inches.	Width b of upper flange in inches.	Sectional area in square inches.	Coefficient K^1.
51	9	18	6.3	31.3	1624
52	10	7	1.1	16.1	672
53	10	8	1.5	17.5	784
54	10	9	2.0	19.0	896
55	10	10	2.4	20.4	1008
56	10	11	2.9	21.9	1106
57	10	12	3.3	23.3	1218
58	10	13	3.8	24.8	1330
59	10	14	4.3	26.3	1428
60	10	15	4.7	27.7	1540
61	10	16	5.2	29.2	1652
62	10	17	5.7	30.7	1750
63	10	18	6.1	32.1	1862
64	11	8	1.3	18.3	896
65	11	9	1.7	19.7	1008
66	11	10	2.2	21.2	1134
67	11	11	2.7	22.7	1246
68	11	12	3.1	24.7	1372
69	11	13	3.6	25.6	1498
70	11	14	4.1	27.1	1610
71	11	15	4.5	28.5	1736
72	11	16	5.0	30.0	1862
73	11	17	5.5	31.5	1974
74	11	18	5.9	32.9	2100
75	12	8	1.1	19.1	994

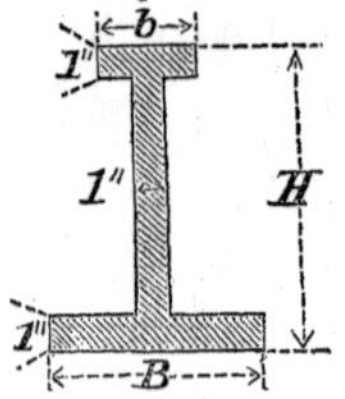

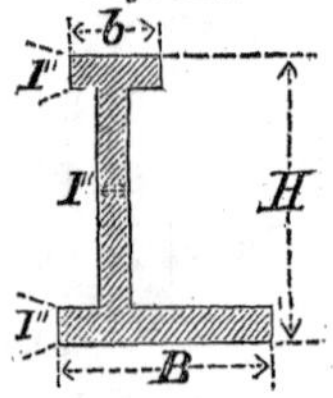

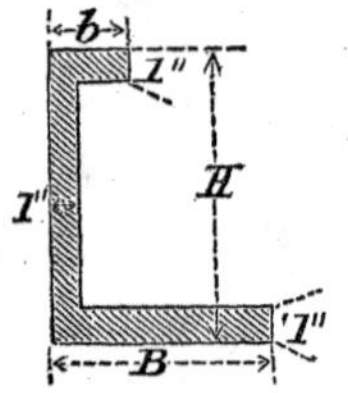

Fig. 146.

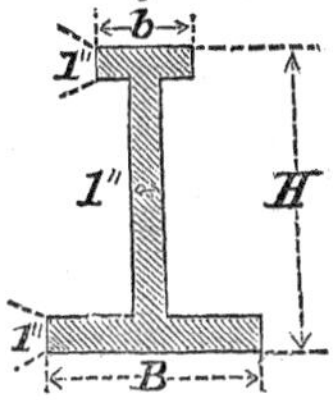

Fig. 147.

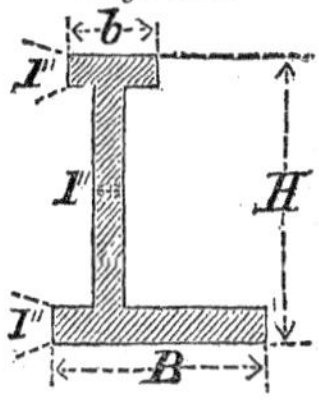

Fig. 148.

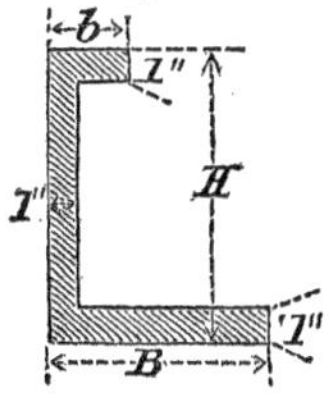

Fig. 149.

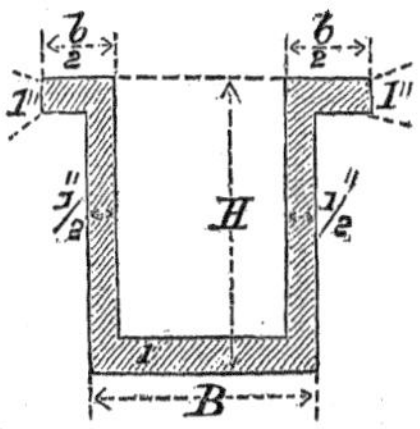

Number of section.	Height H in inches.	Width B of lower flange in inches.	Width b of upper flange in inches.	Sectional area in square inches.	Coefficient $K1$.
76	12	9	1.5	20.5	1120
77	12	10	2.0	22.0	1260
78	12	11	2.5	23.5	1400
79	12	12	2.9	24.9	1526
80	12	13	3.4	26.4	1666
81	12	14	3.9	27.9	1806
82	12	15	4.3	29.3	1932
83	12	16	4.8	30.8	2072
84	12	17	5.3	32.3	2198
85	12	18	5.7	33.7	2338
86	13	9	1.3	21.3	1232
87	13	10	1.8	22.8	1386
88	13	11	2.2	24.2	1540
89	13	12	2.7	25.7	1680
90	13	13	3.2	27.2	1834
91	13	14	3.7	28.7	1988
92	13	15	4.1	30.1	2128
93	13	16	4.6	31.6	2282
94	13	17	5.1	33.1	2422
95	13	18	5.5	34.5	2576
96	14	9	1.1	22.1	1358
97	14	10	1.5	23.5	1512
98	14	11	2.0	25.0	1680
99	14	12	2.5	26.5	1834
100	14	13	3.0	28.0	2002

Fig. 146.

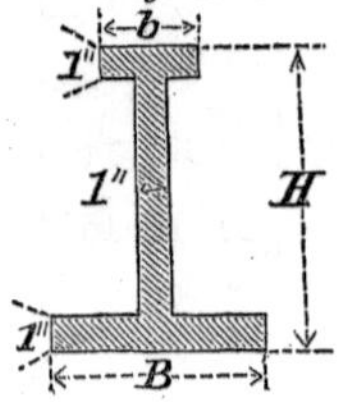

Fig. 147.

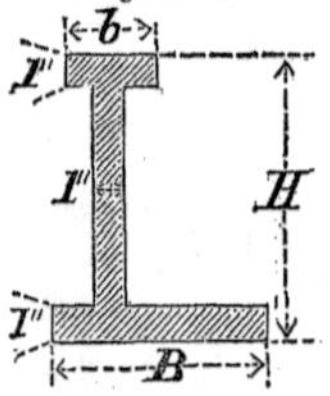

Fig. 148.

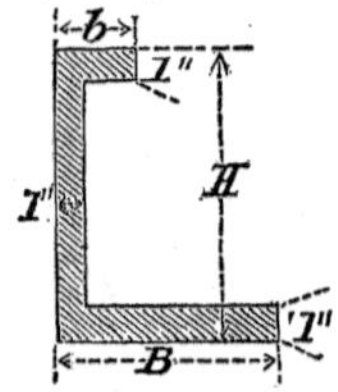

Fig. 149.

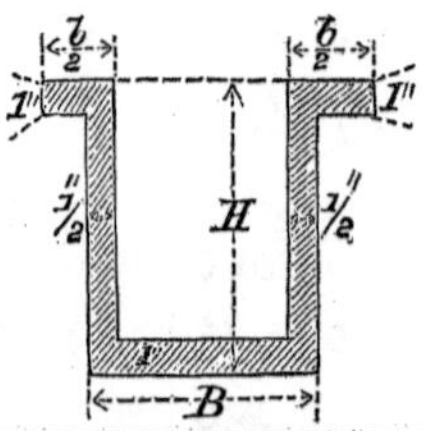

Number of section.	Height H in inches.	Width B of lower flange in inches.	Width b of upper flange in inches.	Sectional area in square inches.	Coefficient K^1.
101	14	14	3.4	29.4	2170
102	14	15	3.9	30.9	2324
103	14	16	4.4	32.4	2492
104	14	17	4.8	33.8	2660
105	14	18	5.3	35.3	2814
106	15	10	1.3	24.3	1638
107	15	11	1.8	25.8	1820
108	15	12	2.3	27.3	2002
109	15	13	2.7	28.7	2170
110	15	14	3.2	30.2	2352
111	15	15	3.7	31.7	2520
112	15	16	4.2	33.2	2702
113	15	17	4.6	34.6	2884
114	15	18	5.1	36.1	3052
115	16	10	1.1	25.1	1764
116	16	11	1.6	26.6	1960
117	16	12	2.0	28.0	2156
118	16	13	2.5	29.5	2338
119	16	14	3.0	31.0	2534
120	16	15	3.5	32.5	2730
121	16	16	3.9	33.9	2912
122	16	17	4.4	35.4	3108
123	16	18	4.9	36.9	3290
124	17	11	1.3	27.3	2100
125	17	12	1.8	28.8	2310

Fig. 146.

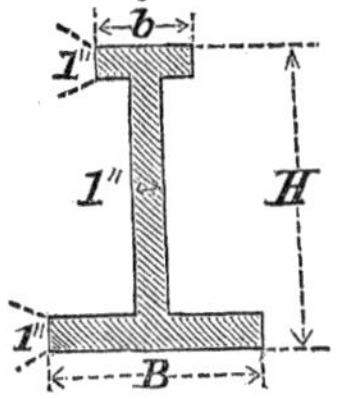

Fig. 147.

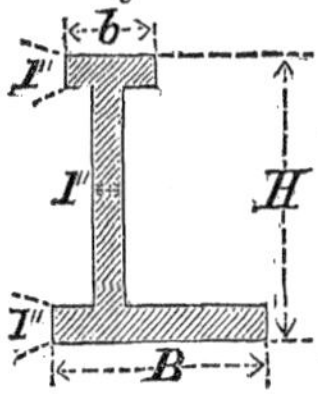

Fig. 148.

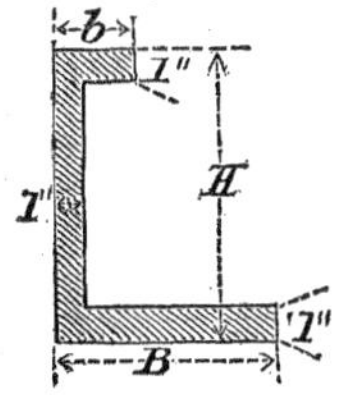

Fig. 149.

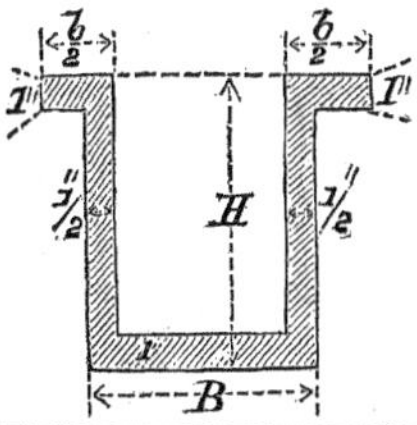

Number of section.	Height H in inches.	Width B of lower flange in inches.	Width b of upper flange in inches.	Sectional area in square inches.	Coefficient K^1.
126	17	13	2.3	30.3	2506
127	17	14	2.8	31.8	2716
128	17	15	3.2	33.2	2926
129	17	16	3.7	34.7	3122
130	17	17	4.2	36.2	3332
131	17	18	4.7	37.7	3542
132	18	11	1.1	28.1	2240
133	18	12	1.6	29.6	2464
134	18	13	2.0	31.0	2688
135	18	14	2.5	32.5	2898
136	18	15	3.0	34.0	3122
137	18	16	3.5	35.5	3346
138	18	17	4.0	37.0	3556
139	18	18	4.4	38.4	3780

Fig. 150.

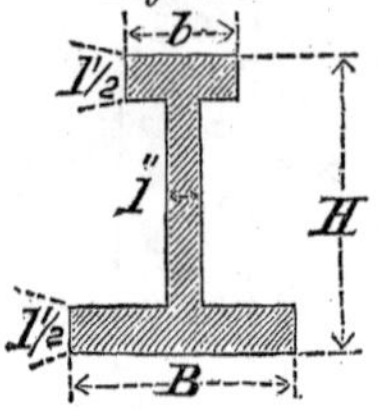

Fig. 151.

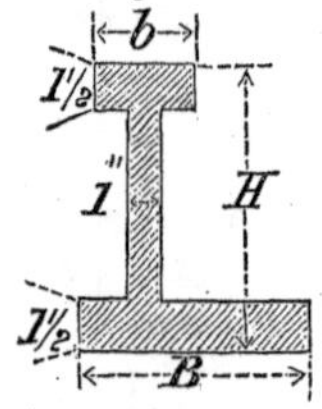

Fig. 152.

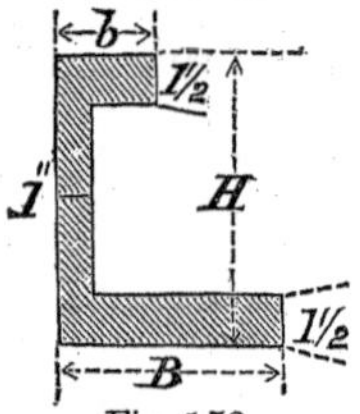

Fig. 153.

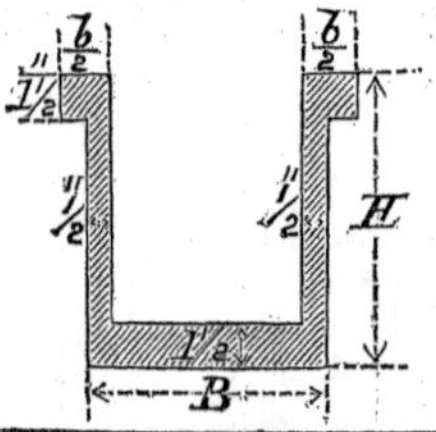

Number of section.	Height H in inches.	Width B of lower flange in inches.	Width b of upper flange in inches.	Sectional area in square inches.	Coefficient K^1.
140	6	5	1.3	12.5	280
141	6	6	1.7	14.6	336
142	6	7	2.1	16.7	406
143	6	8	2.5	18.8	462
144	6	9	2.9	20.9	518
145	6	10	3.2	22.8	588
146	6	11	3.6	24.9	644
147	6	12	4.0	27.0	714
148	6	13	4.4	29.1	770
149	6	14	4.8	31.2	826
150	6	15	5.2	33.3	896
151	6	16	5.5	35.3	952
152	6	17	5.9	37.4	1022
153	6	18	6.3	39.5	1078
154	7	5	1.2	13.3	364
155	7	6	1.6	15.4	434
156	7	7	2.0	17.5	518
157	7	8	2.4	19.6	602
158	7	9	2.8	21.7	686
159	7	10	3.2	23.8	756
160	7	11	3.7	26.1	840
161	7	12	4.1	28.2	924
162	7	13	4.5	30.3	1008
163	7	14	4.9	32.4	1092
164	7	15	5.3	34.5	1162

Fig. 150.

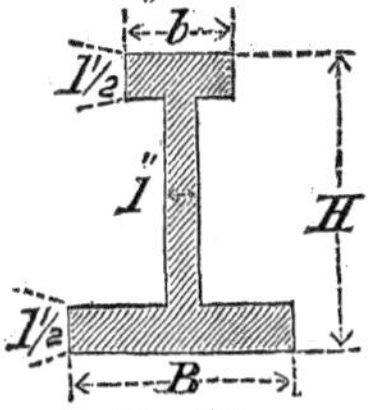

Fig. 151.

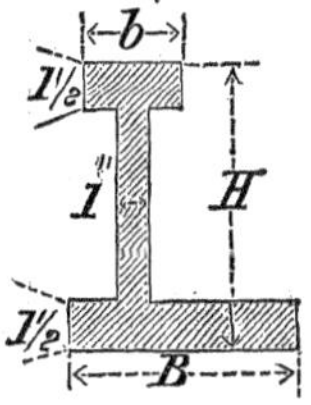

Fig. 152.

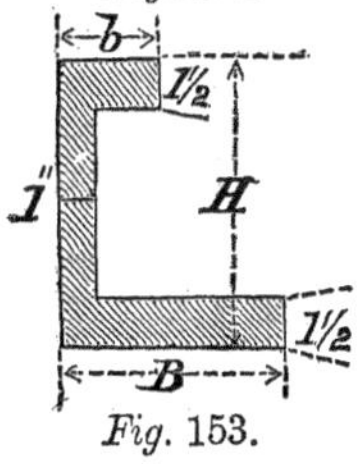

Fig. 153.

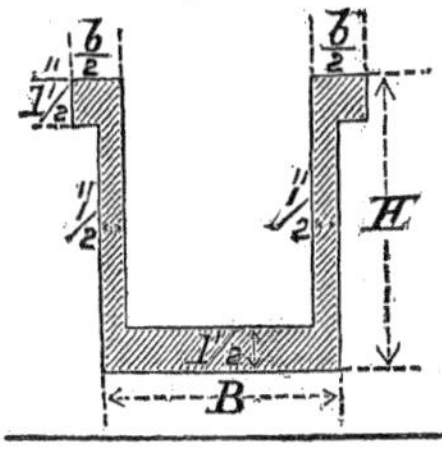

Number of section.	Height H in inches.	Width B of lower flange in inches.	Width b of upper flange in inches.	Sectional area in square inches.	Coefficient K^1.
165	7	16	5.7	36.6	1246
166	7	17	6.1	38.7	1330
167	7	18	6.5	40.8	1414
168	8	5	1.1	14.2	1022
169	8	6	1.5	16.3	546
170	8	7	2.0	18.5	644
171	8	8	2.4	20.6	742
172	8	9	2.8	22.7	840
173	8	10	3.2	24.8	938
174	8	11	3.6	26.9	1036
175	8	12	4.1	29.2	1148
176	8	13	4.5	31.3	1246
177	8	14	4.9	33.4	1344
178	8	15	5.3	35.5	1442
179	8	16	5.7	37.6	1540
180	8	17	6.2	39.8	1638
181	8	18	6.6	41.9	1750
182	9	5	1.0	15.0	518
183	9	6	1.4	17.1	644
184	9	7	1.9	19.4	770
185	9	8	2.3	21.5	882
186	9	9	2.7	23.6	1008
187	9	10	3.1	25.7	1120
188	9	11	3.6	27.9	1246
189	9	12	4.0	30.0	1358

Fig. 150.

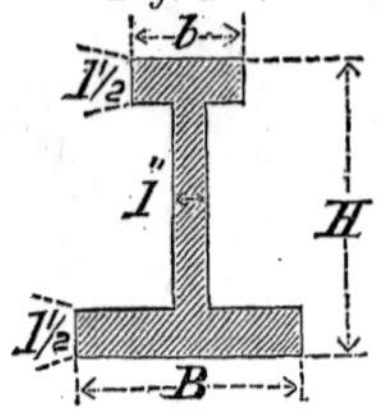

Fig. 151.

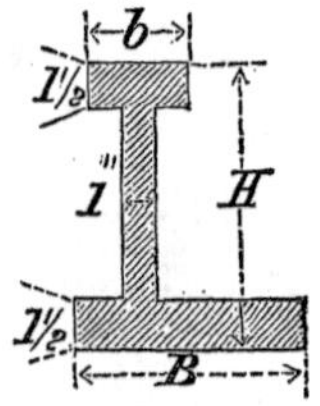

Fig. 152.

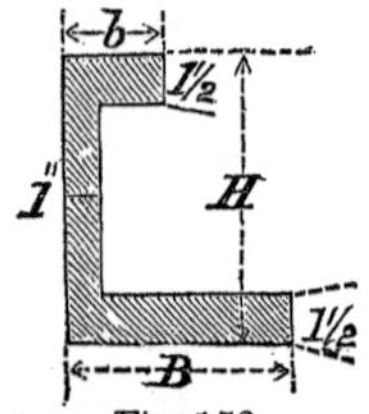

Fig. 153.

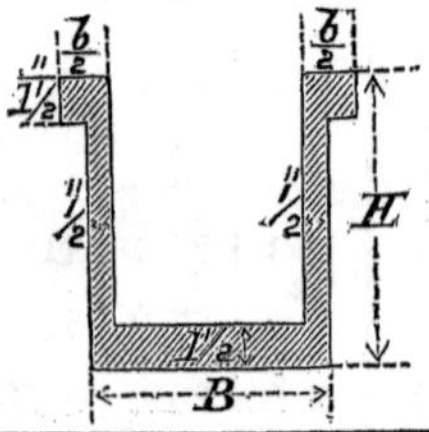

Number of section.	Height H in inches.	Width B of lower flange in inches.	Width b of upper flange in inches.	Sectional area in square inches.	Coefficient K^1.
190	9	13	4.4	32.1	1484
191	9	14	4.9	34.4	1610
192	9	15	5.3	36.5	1722
193	9	16	5.7	38.6	1848
194	9	17	6.2	40.8	1960
195	9	18	6.6	42.9	2086
196	10	6	1.3	18.0	756
197	10	7	1.7	20.1	896
198	10	8	2.2	22.3	1036
199	10	9	2.6	24.4	1176
200	10	10	3.1	26.7	1316
201	10	11	3.5	28.8	1456
202	10	12	3.9	30.9	1596
203	10	13	4.4	33.1	1736
204	10	14	4.8	35.2	1876
205	10	15	5.2	37.3	2016
206	10	16	5.7	39.6	2156
207	10	17	6.1	41.7	2296
208	10	18	6.5	43.8	2436
209	11	6	1.2	18.8	854
210	11	7	1.6	20.9	1022
211	11	8	2.1	23.2	1176
212	11	9	2.5	25.3	1344
213	11	10	3.0	27.5	1498
214	11	11	3.4	29.6	1666

Fig. 150.

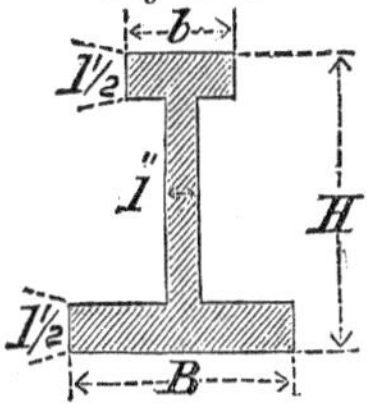

Fig. 151.

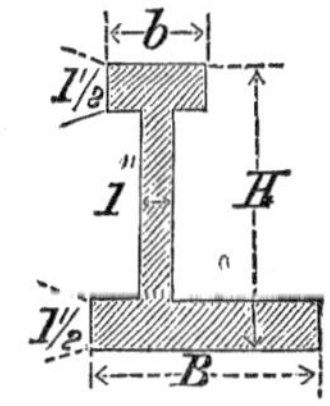

Fig. 152.

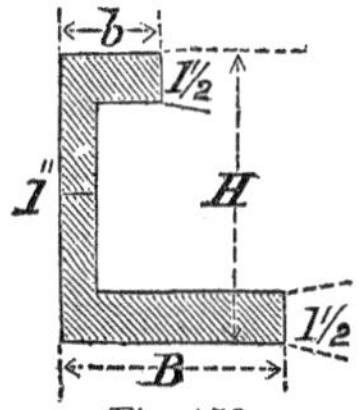

Fig. 153.

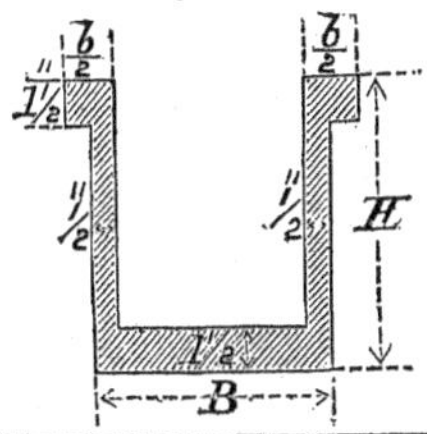

Number of section.	Height H in inches.	Width B of lower flange in inches.	Width b of upper flange in inches.	Sectional area in square inches.	Coefficient K^1.
215	11	12	3.8	31.7	1820
216	11	13	4.3	34 0	1974
217	11	14	4.7	36.1	2128
218	11	15	5.2	38.3	2296
219	11	16	5.6	40.4	2464
220	11	17	6.1	42.7	2618
221	11	18	6.5	44.8	2786
222	12	6	1.0	19.5	966
223	12	7	1.5	21.8	1148
224	12	8	1.9	23.9	1330
225	12	9	2.4	26.1	1512
226	12	10	2.8	28.2	1680
227	12	11	3.3	30.5	1862
228	12	12	3.7	32.6	2044
229	12	13	4.2	34.8	2226
230	12	14	4.6	36.9	2408
231	12	15	5.1	39.2	2590
232	12	16	5.5	41.3	2772
233	12	17	6.0	43.5	2954
234	12	18	6.4	45.6	3136
235	13	7	1.4	22.6	1274
236	13	8	1.8	24.7	1470
237	13	9	2.3	27.0	1680
238	13	10	2.7	29.1	1876
239	13	11	3.2	31.3	2072

Fig. 150.

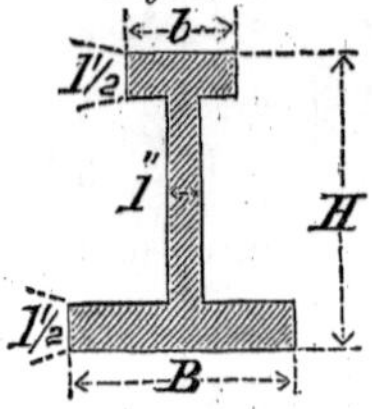

Fig. 151.

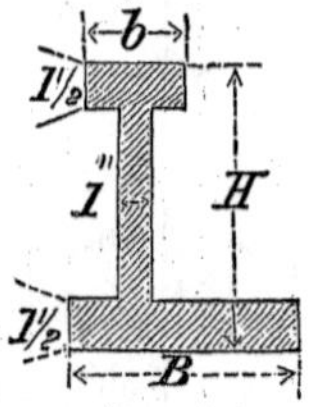

Fig. 152.

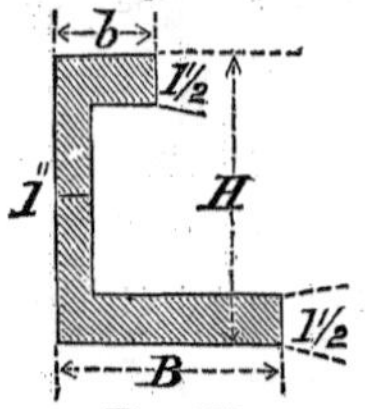

Fig. 153.

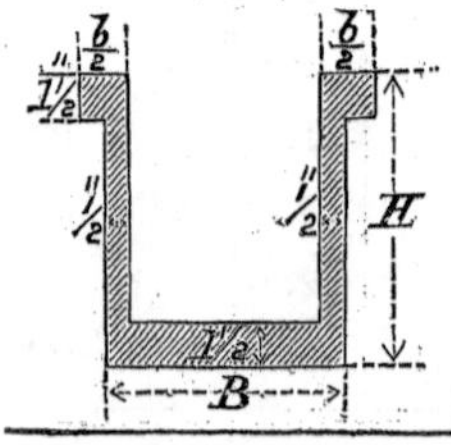

Number of section.	Height H in inches.	Width B of lower flange in inches.	Width b of upper flange in inches.	Sectional area in square inches.	Coefficient K^1.
240	13	12	3.6	33.4	2282
241	13	13	4.1	35.7	2478
242	13	14	4.5	37.8	2674
243	13	15	5.0	40.0	2884
244	13	16	5.4	42.1	3080
245	13	17	5.9	44.4	3276
246	13	18	6.3	46.5	3486
247	14	7	1.2	23.3	1400
248	14	8	1.7	25.6	1624
249	14	9	2.1	27.7	1848
250	14	10	2.6	29.9	2058
251	14	11	3.0	32.0	2282
252	14	12	3.5	34.3	2506
253	14	13	3.9	36.4	2730
254	14	14	4.4	38.6	2954
255	14	15	4.9	40.9	3178
256	14	16	5.3	43.0	3388
257	14	17	5.8	45.2	3612
258	14	18	6.2	47.3	3836
259	15	7	1.1	24.2	1526
260	15	8	1.5	26.3	1764
261	15	9	2.0	28.5	2016
262	15	10	2.4	30.6	2254
263	15	11	2.9	32.9	2492
264	15	12	3.4	35.1	2744

Fig. 150.

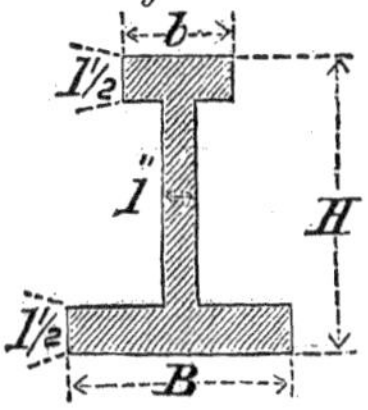

Fig. 151.

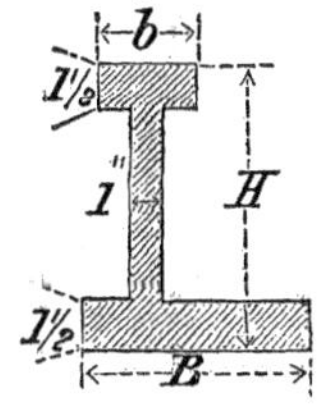

Fig. 152.

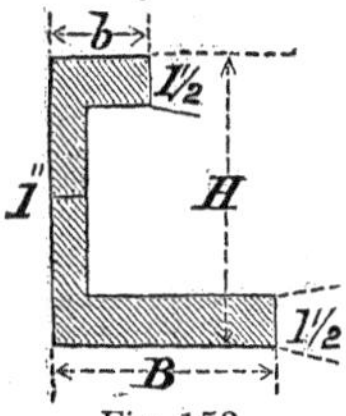

Fig. 153.

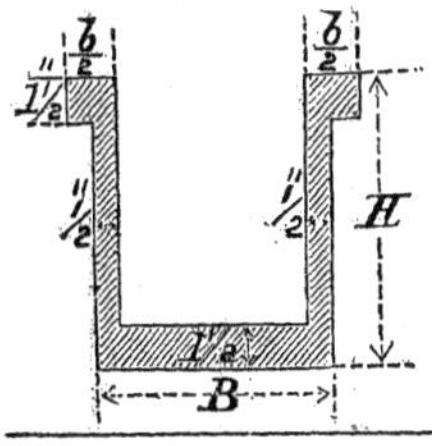

Number of section.	Height H in inches.	Width B of lower flange in inches.	Width b of upper flange in inches.	Sectional area in square inches.	Coefficient $K1$.
265	15	13	3.8	37.2	2982
266	15	14	4.3	39.5	3220
267	15	15	4.7	41.6	3472
268	15	16	5.2	43.8	3710
269	15	17	5.7	46.1	3948
270	15	18	6.1	48.2	4200
271	16	8	1 4	27.1	1918
272	16	9	1.8	29.2	2184
273	16	10	2.3	31.5	2450
274	16	11	2.8	33.7	2702
275	16	12	3.2	35.8	2968
276	16	13	3.7	38.1	3234
277	16	14	4.1	40.2	3500
278	16	15	4.7	42.6	3766
279	16	16	5.2	44.8	4018
280	16	17	5.7	47.1	4284
281	16	18	6.1	49.2	4550
282	17	8	1.2	27.8	2072
283	17	9	1.7	30.1	2352
284	17	10	2.1	32.2	2632
285	17	11	2.6	34.4	2926
286	17	12	3.1	36.7	3206
287	17	13	3.5	38.8	3486
288	17	14	4.0	41.0	3766
289	17	15	4.5	43.3	4060

Fig. 150.

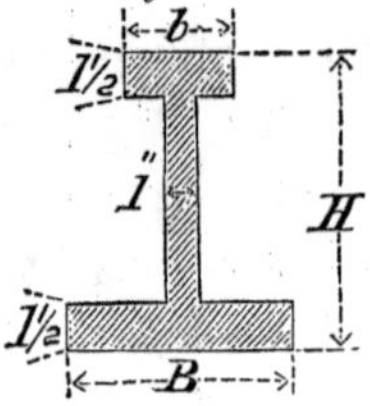

Fig. 151.

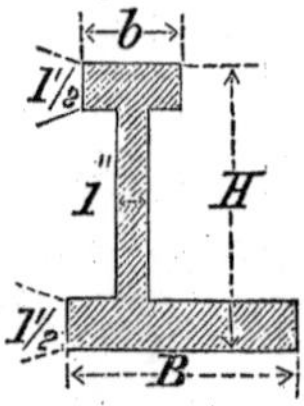

Fig. 152.

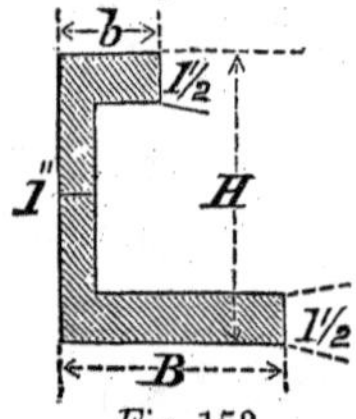

Fig. 153.

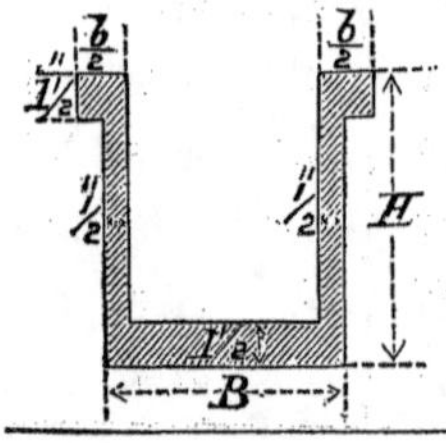

Number of section.	Height H in inches.	Width B of lower flange in inches.	Width b of upper flange in inches.	Sectional area in square inches.	Coefficient K^1.
290	17	16	4.9	45.4	4340
291	17	17	5.4	47.6	4620
292	17	18	5.9	49.9	4900
293	18	8	1.1	28.7	2226
294	18	9	1.5	30.8	2520
295	18	10	2.0	33.0	2828
296	18	11	2.5	35.3	3136
297	18	12	2.9	37.4	3430
298	18	13	3.4	39.6	3738
299	18	14	3.9	41.9	4056
300	18	15	4.3	44.0	4354
301	18	16	4.8	46.2	4648
302	18	17	5.3	48.5	4956
303	18	18	5.7	50.6	5269

Fig. 154.

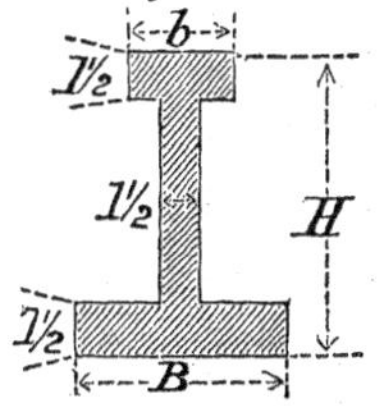

Fig. 155.

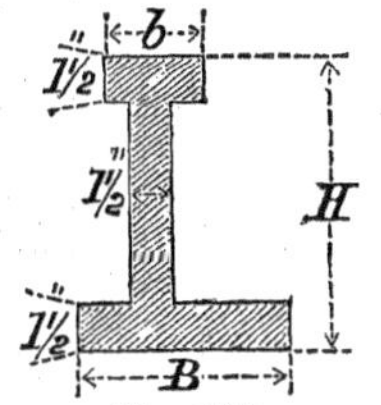

Fig. 156.

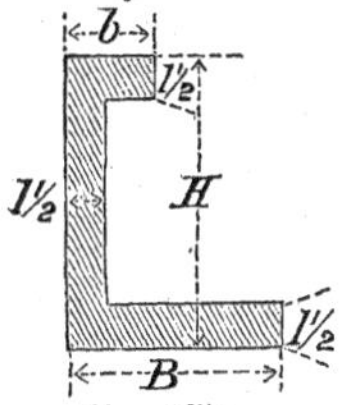

Fig. 157.

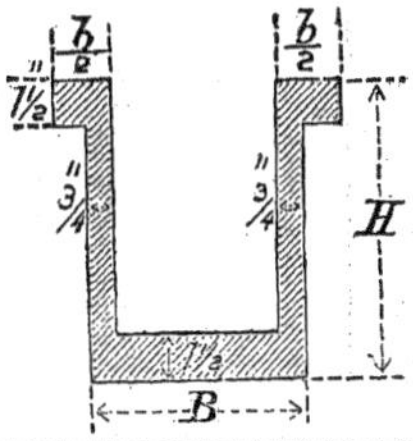

Number of section.	Height H in inches.	Width B of lower flange in inches.	Width b of upper flange in inches.	Sectional area in square inches.	Coefficient $K1$.
304	6	7	1.8	17.7	378
305	6	8	2.2	19.8	448
306	6	9	2.5	21.8	504
307	6	10	2.9	23.9	574
308	6	11	3.3	26.0	630
309	6	12	3.7	28.1	686
310	6	13	4.1	30.2	756
311	6	14	4.5	32.3	812
312	6	15	4.9	34.4	882
313	6	16	5.2	36.3	938
314	6	17	5.6	38.4	1008
315	6	18	6.0	40.5	1064
316	7	7	1.6	18.9	490
317	7	8	2.0	21.0	574
318	7	9	2.4	23.1	658
319	7	10	2.8	25.2	742
320	7	11	3.3	27.5	826
321	7	12	3.7	29.6	896
322	7	13	4.1	31.7	980
323	7	14	4.5	33.8	1064
324	7	15	4.9	35.9	1148
325	7	16	5.3	38.0	1232
326	7	17	5.7	40.1	1302
327	7	18	6.1	42.2	1386
328	8	8	1.9	22.4	714

Fig. 154.

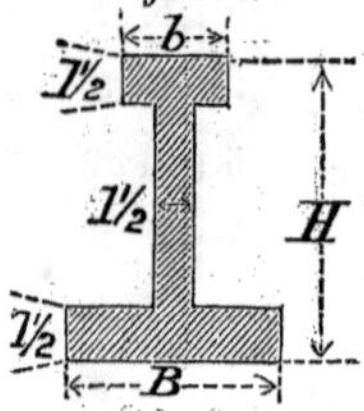

Fig. 155.

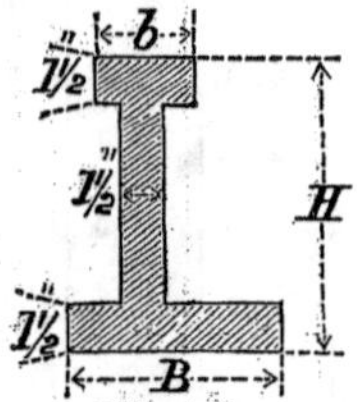

Fig. 156.

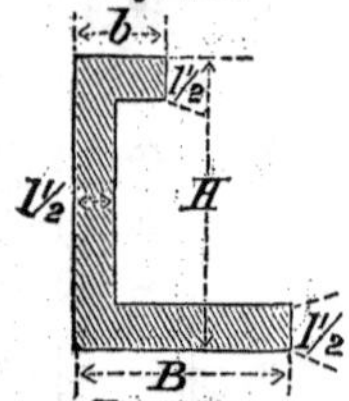

Fig. 157.

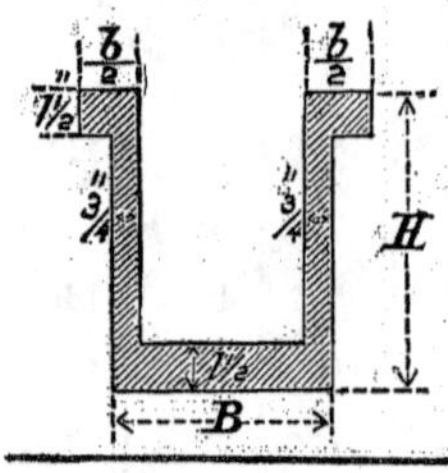

Number of section.	Height H in inches.	Width B of lower flange in inches.	Width b of upper flange in inches.	Sectional area in square inches.	Coefficient K^1.
329	8	9	2.3	24.5	812
330	8	10	2.7	26.6	910
331	8	11	3.1	28.7	1008
332	8	12	3.6	30.9	1106
333	8	13	4.0	33.0	1218
334	8	14	4.4	35.1	1316
335	8	15	4.8	37.2	1414
336	8	16	5.2	39.3	1512
337	8	17	5.7	41.6	1610
338	8	18	6.1	43.7	1708
339	9	8	1.7	23.6	840
340	9	9	2.1	25.7	966
341	9	10	2.6	27.9	1092
342	9	11	3.0	30.0	1204
343	9	12	3.4	32.1	1330
344	9	13	3.9	34.4	1442
345	9	14	4.3	36.5	1568
346	9	15	4.7	38.6	1694
347	9	16	5.1	40.7	1806
348	9	17	5.6	42.9	1932
349	9	18	6.0	45.0	2044
350	10	8	1.5	24.8	980
351	10	9	2.0	27.0	1120
352	10	10	2.4	29.1	1260
353	10	11	2.8	31.2	1400

Fig. 154.

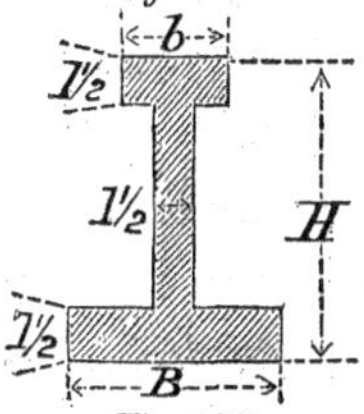

Fig. 155.

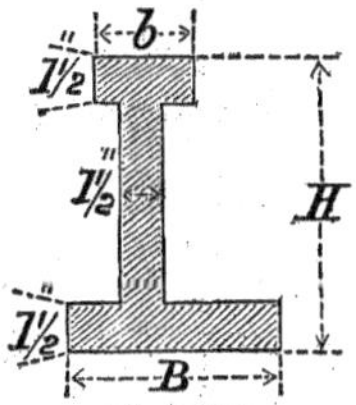

Fig. 156.

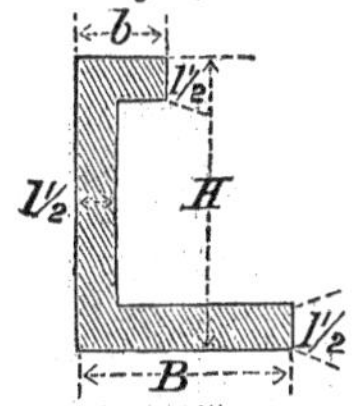

Fig. 157.

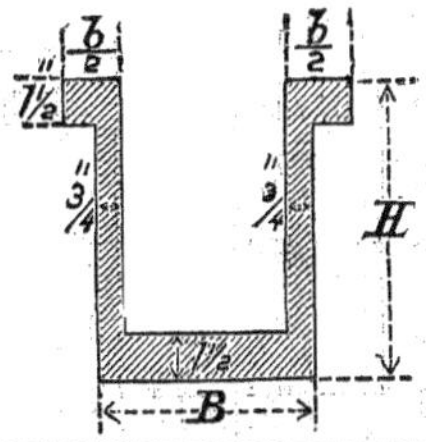

Number of section.	Height H in inches.	Width B of lower flange in inches.	Width b of upper flange in inches.	Sectional area in square inches.	Coefficient K^1.
354	10	12	3.3	33.5	1540
355	10	13	3.7	35.6	1680
356	10	14	4.1	37.7	1820
357	10	15	4.6	39.9	1960
358	10	16	5.0	42.0	2100
359	10	17	5.5	44.3	2240
360	10	18	5.9	46.4	2380
361	11	9	1.8	28.2	1288
362	11	10	2.2	30.3	1442
363	11	11	2.6	32.4	1610
364	11	12	3.1	34.7	1764
365	11	13	3.5	36.8	1932
366	11	14	4.0	39.0	2086
367	11	15	4.4	41.1	2240
368	11	16	4.9	43.4	2408
369	11	17	5.3	45.5	2562
370	11	18	5.8	47.7	2730
371	12	9	1.6	29.4	1442
372	12	10	2.0	31.5	1624
373	12	11	2.5	33.8	1806
374	12	12	2.9	35.9	1988
375	12	13	3.4	38.1	2170
376	12	14	3.8	40.2	2352
377	12	15	4.2	42.3	2534
378	12	16	4.7	44.6	2716

Fig. 154.

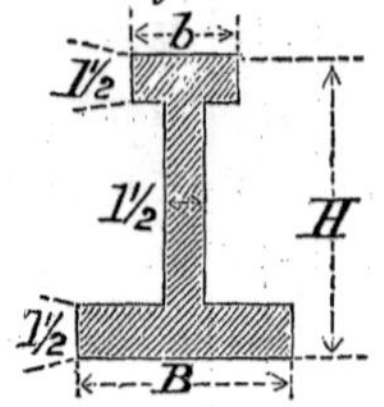

Fig. 155.

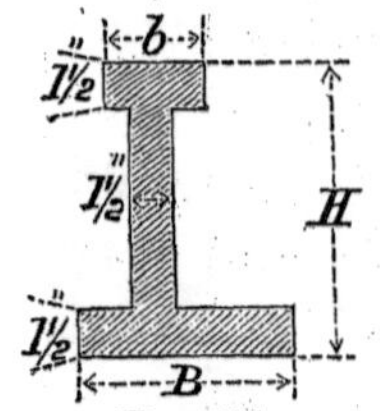

Fig. 156.

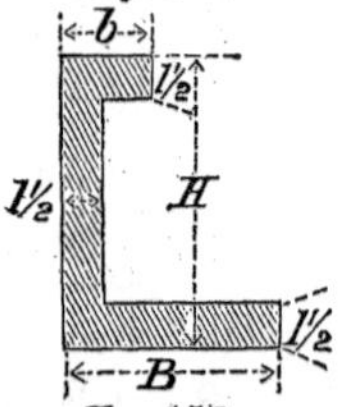

Fig. 157.

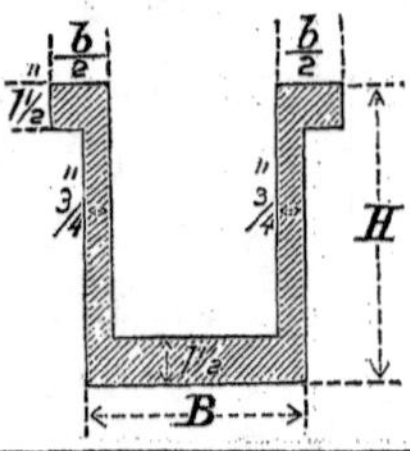

Number of section.	Height H in inches.	Width B of lower flange in inches.	Width b of upper flange in inches.	Sectional area in square inches.	Coefficient K^1.
379	12	17	5.1	46.7	2898
380	12	18	5.6	48.9	3066
381	13	10	1.8	32.7	1806
382	13	11	2.2	34.8	2002
383	13	12	2.7	37.1	2212
384	13	13	3.2	39.3	2408
385	13	14	3.6	41.4	2618
386	13	15	4.1	43.7	2814
387	13	16	4.5	45.8	3010
388	13	17	5.0	48.0	3220
389	13	18	5.4	50.1	3416
390	14	10	1.6	33.9	1988
391	14	11	2.0	36.0	2212
392	14	12	2.5	38.3	2436
393	14	13	2.9	40.4	2660
394	14	14	3.4	42.6	2870
395	14	15	3.9	44.9	3094
396	14	16	4.3	47.0	3318
397	14	17	4.8	49.2	3542
398	14	18	5.2	51.3	3766
399	15	11	1.8	37.2	2408
400	15	12	3.3	39.7	2660
401	15	13	2.7	41.6	2898
402	15	14	3.2	43.8	3136
403	15	15	3.7	46.1	3388

Fig. 154.

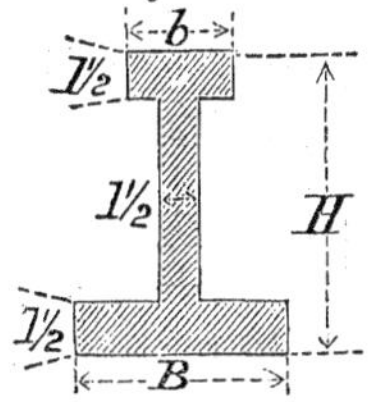

Fig. 155.

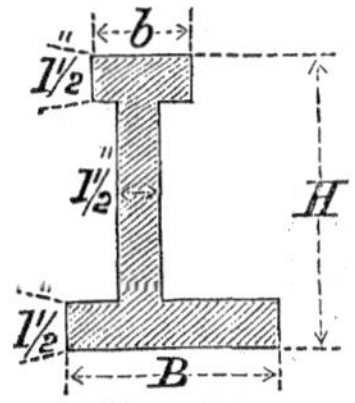

Fig. 156.

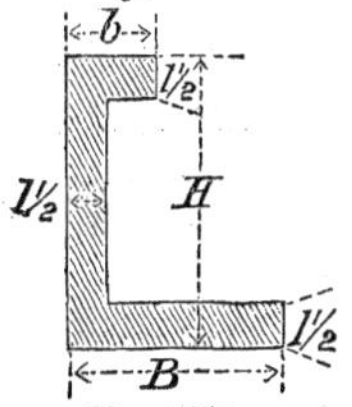

Fig. 157.

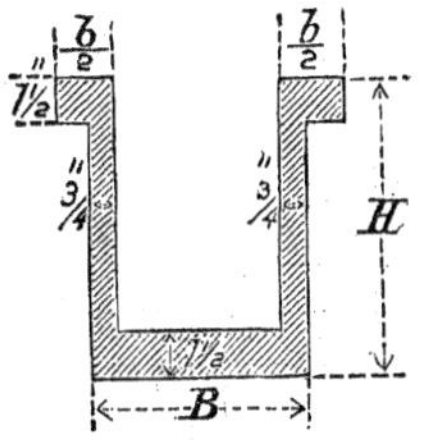

Number of section.	Height *H* in inches.	Width *B* of lower flange in inches.	Width *b* of upper flange in inches.	Sectional area in square inches.	Coefficient K^1.
404	15	16	4.1	48.2	3626
405	15	17	4.6	50.4	3864
406	15	18	5.0	52.5	4116
407	16	11	1.6	38.4	2618
408	16	12	2.1	40.7	2884
409	16	13	2.5	42.8	3136
410	16	14	3.0	45.0	3402
411	16	15	3.4	47.1	3668
412	16	16	3.9	49.4	3934
413	16	17	4.4	51.6	4186
414	16	18	4.8	53.7	4452
415	17	12	1.8	41.7	3108
416	17	13	2.3	44.0	3388
417	17	14	2.8	46.2	3682
418	17	15	3.2	48.3	3962
419	17	16	3.7	50.6	4242
420	17	17	4.2	52.8	4522
421	17	18	4.6	54.9	4816
422	18	12	1.6	42.9	3332
423	18	13	2.1	45.2	3626
424	18	14	2.5	47.3	3934
425	18	15	3.0	49.5	4242
426	18	16	3.5	51.8	4550
427	18	17	3.9	53.9	4858
428	18	18	4.4	56.1	5152

Fig. 158.

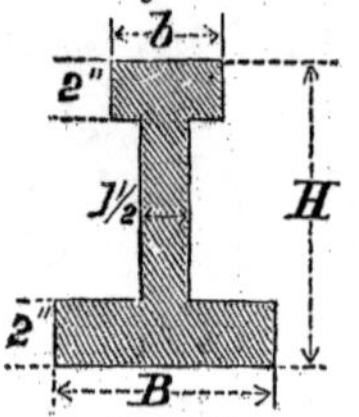

Fig. 159.

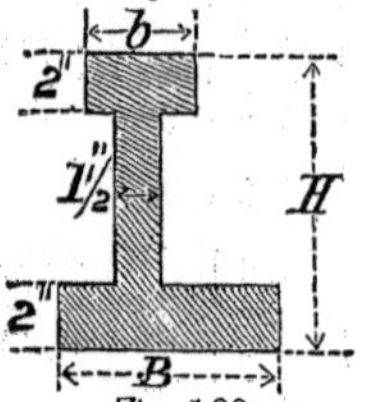

Fig. 160.

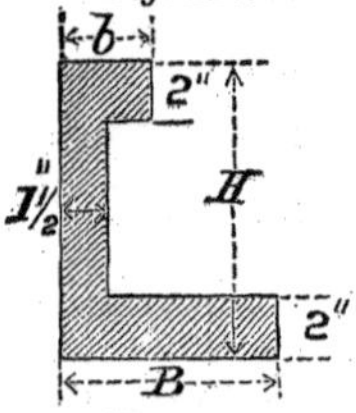

Fig. 161.

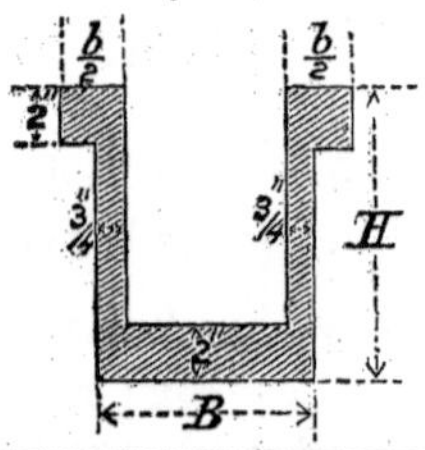

Number of section.	Height H in inches.	Width B of lower flange in inches.	Width b of upper flange in inches.	Sectional area in square inches.	Coefficient K^1.
429	6	6	1.5	18.0	336
430	6	7	1.8	20.6	392
431	6	8	2.2	23.4	462
432	6	9	2.5	26.0	518
433	6	10	2.8	28.6	588
434	6	11	3.2	31.4	624
435	6	12	3.5	34.0	714
436	6	13	3.8	36.6	770
437	6	14	4.2	39.4	840
438	6	15	4.5	42.0	896
439	6	16	4.8	44.6	952
440	6	17	5.2	47.4	1022
441	6	18	5.5	50.0	1078
442	7	7	1.8	22.1	532
443	7	8	2.2	24.9	616
444	7	9	2.6	27.7	714
445	7	10	2.9	30.3	798
446	7	11	3.3	33.1	882
447	7	12	3.7	35.9	966
448	7	13	4.0	38.5	1050
449	7	14	4.4	41.3	1134
450	7	15	4.7	43.9	1218
451	7	16	5.1	46.7	1302
452	7	17	5.5	49.5	1386
453	7	18	5.8	52.1	1470

Fig. 158.

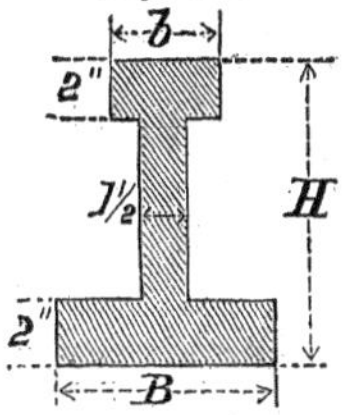

Fig. 159.

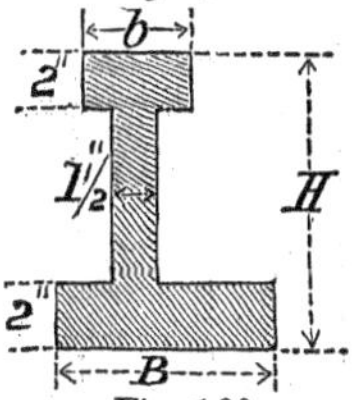

Fig. 160.

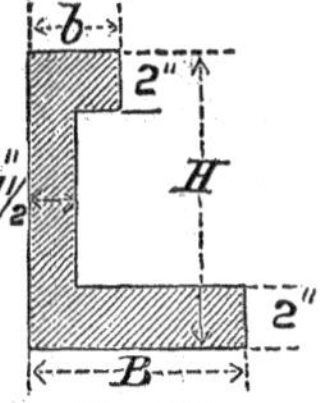

Fig. 161.

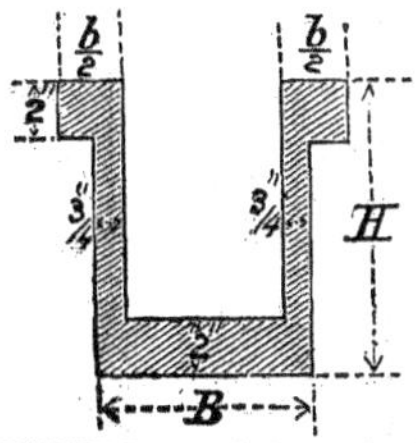

Number of section.	Height H in inches.	Width B of lower flange in inches.	Width b of upper flange in inches.	Sectional area in square inches.	Coefficient $K1$.
454	8	7	1.8	23.6	686
455	8	8	2.2	26.4	714
456	8	9	2.5	29.0	896
457	8	10	2.9	31.8	1008
458	8	11	3.3	34.6	1120
459	8	12	3.7	37.4	1232
460	8	13	4.1	40.2	1344
461	8	14	4.5	43.0	1456
462	8	15	4.9	45.8	1554
463	8	16	5.2	48 4	1666
464	8	17	5.6	51.2	1778
465	8	18	6.0	54.0	1890
466	9	7	1.7	24.9	826
467	9	8	2.1	27.7	966
468	9	9	2.5	30.5	1106
469	9	10	2.9	33.3	1232
470	9	11	3.3	36.1	1372
471	9	12	3.7	38.9	1498
472	9	13	4.1	41.7	1638
473	9	14	4.5	44.5	1778
474	9	15	4.9	47.3	1904
475	9	16	5.3	50.1	2044
476	9	17	5.7	52.9	2184
477	9	18	6.1	55.7	2310
478	10	7	1.6	26.2	980

Fig. 158.

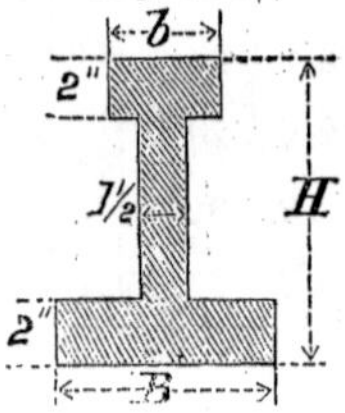

Fig. 159.

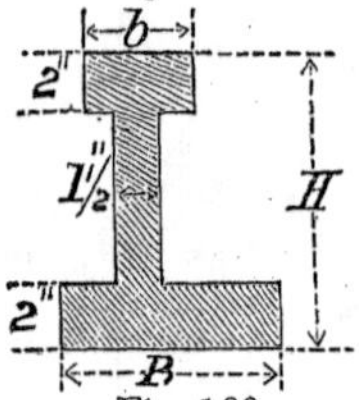

Fig. 160.

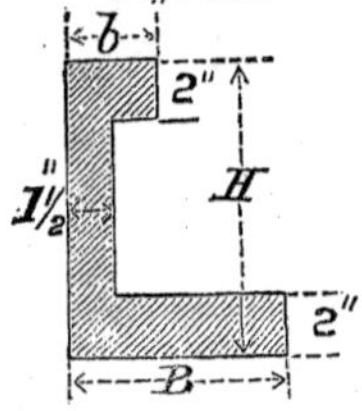

Fig. 161.

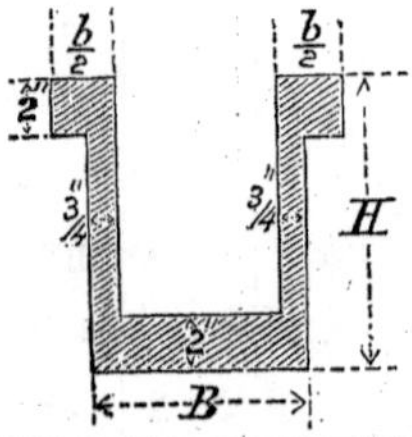

Number of section.	Height H in inches.	Width B of lower flange in inches.	Width b of upper flange in inches.	Sectional area in square inches.	Coefficient $K1$.
479	10	8	2.0	29.0	1134
480	10	9	2.4	31.8	1302
481	10	10	2.8	34.6	1456
482	10	11	3.2	37.4	1624
483	10	12	3.6	40.2	1778
484	10	13	4.0	43.0	1946
485	10	14	4.4	45.8	2100
486	10	15	4.9	48.8	2268
487	10	16	5.3	51.6	2422
488	10	17	5.7	54.4	2590
489	10	18	6.1	57.2	2744
490	11	8	1.9	30.3	1316
491	11	9	2.3	33.1	1512
492	11	10	2.7	35.9	1694
493	11	11	3.1	38.7	1876
494	11	12	3.5	41.5	2072
495	11	13	4.0	44.5	2254
496	11	14	4.4	47.3	2436
497	11	15	4.8	50.1	2632
498	11	16	5.2	52.9	2814
499	11	17	5.6	55.7	2996
500	11	18	6.1	58.7	3192
501	12	8	1.7	31.4	1512
502	12	9	2.1	34.2	1722
503	12	10	2.6	37.2	1932

Fig. 158.

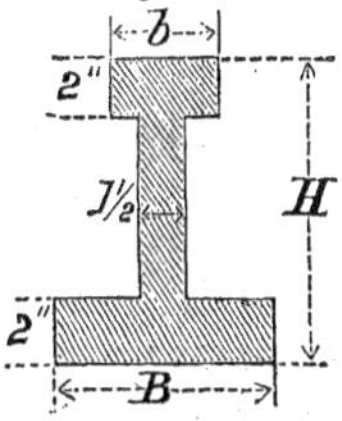

Fig. 159.

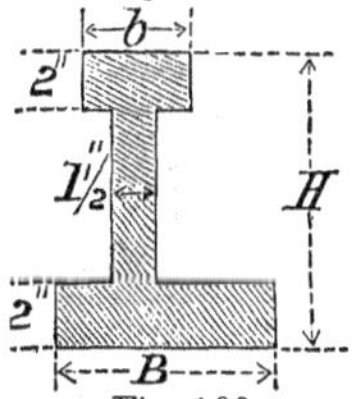

Fig. 160.

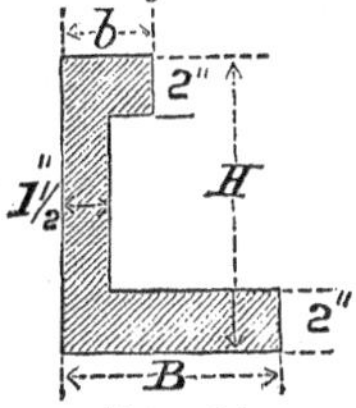

Fig. 161.

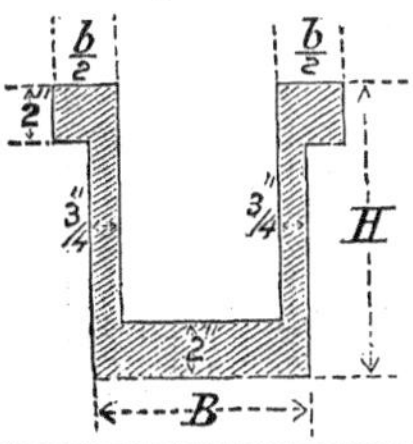

Number of section.	Height H in inches.	Width B of lower flange in inches.	Width b of upper flange in inches.	Sectional area in square inches.	Coefficient K^1.
504	12	11	3.0	40.0	2142
505	12	12	3.4	42.8	2366
506	12	13	3 9	45.8	2576
507	12	14	4.3	48.6	2786
508	12	15	4.7	51.4	2996
509	12	16	5.2	54.4	3220
510	12	17	5.6	57 2	3430
511	12	18	6.0	60 0	3640
512	13	8	1.6	32.7	1680
513	13	9	2.0	35.5	1932
514	13	10	2.4	38.3	2170
515	13	11	2.9	41 3	2408
516	13	12	3.3	44.1	2646
517	13	13	3.8	47.1	2884
518	13	14	4.2	49.9	3122
519	13	15	4.6	52.7	3360
520	13	16	5.1	55.7	3598
521	13	17	5.5	58.5	3850
522	13	18	5.9	61.3	4088
523	14	9	1.9	36.8	2142
524	14	10	2.3	39.6	2408
525	14	11	2.7	42.4	2674
526	14	12	3.2	45.4	2940
527	14	13	3.6	48.2	3206
528	14	14	4.1	52.2	3472

Fig. 158.

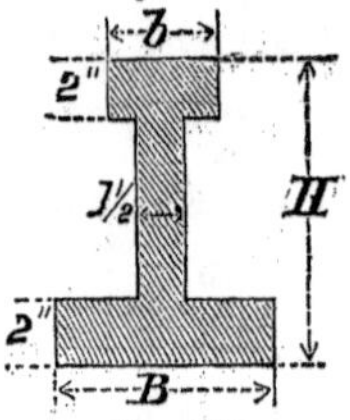

Fig. 159.

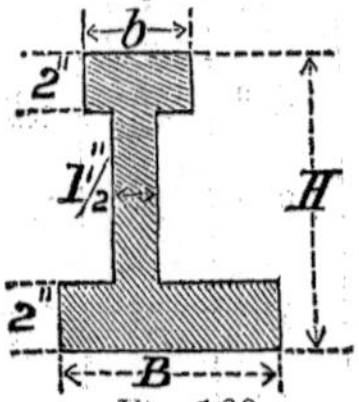

Fig. 160.

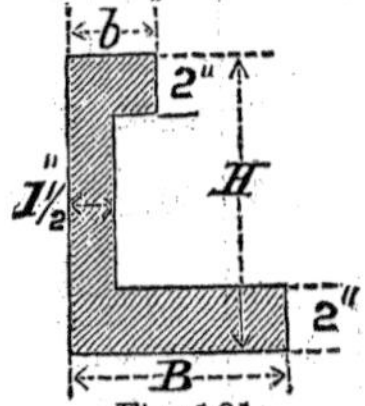

Fig. 161.

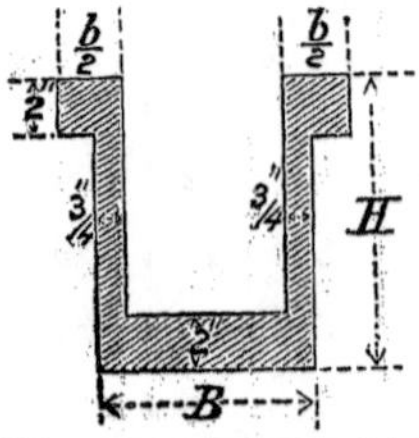

Number of section.	Height H in inches.	Width B of lower flange in inches.	Width b of upper flange in inches.	Sectional area in square inches.	Coefficient K^1.
529	14	15	4.5	54.0	3738
530	14	16	4.9	56.8	4004
531	14	17	5.4	59.8	4270
532	14	18	5.8	62.6	4536
533	15	9	1.7	37.9	2352
534	15	10	2.2	40.9	2646
535	15	11	2.6	43.7	2940
536	15	12	3.0	46.5	3234
537	15	13	3.5	49.5	3528
538	15	14	3.9	52.3	3822
539	15	15	4.4	55.3	4116
540	15	16	4.8	58.1	4410
541	15	17	5.3	61.1	4704
542	15	18	5.7	63.9	4998
543	16	9	1.6	39.2	2562
544	16	10	2.0	42.0	2884
545	16	11	2.5	45.0	3206
546	16	12	2.9	47.8	3528
547	16	13	3.4	50.8	3850
548	16	14	3.8	53.6	4172
549	16	15	4.3	56.6	4494
550	16	16	4.7	59.4	4816
551	16	17	5.2	65.4	5138
551	16	18	5.6	62.2	5460
552	17	10	1.9	43.3	3150

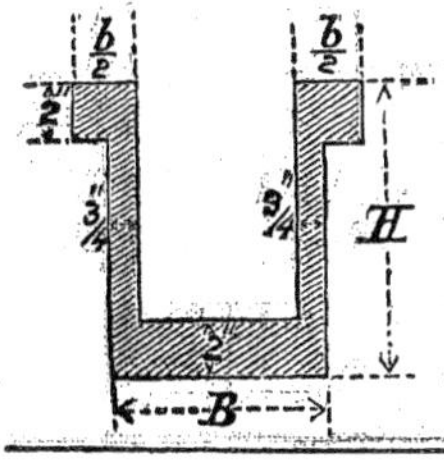

	Number of section.	Height H in inches.	Width B of lower flange in inches.	Width b of upper flange in inches.	Sectional area in square inches.	Coefficient K^1.
Fig. 158.	554	17	11	2.3	46.1	3486
	555	17	12	2.8	49.1	3836
	556	17	13	3.2	51.9	4186
	557	17	14	3.7	54.9	4536
Fig. 159.	558	17	15	4.1	57.7	4872
	559	17	16	4.6	60.7	5222
	560	17	17	5.0	63.5	5572
	561	17	18	5.5	66.5	5922
Fig. 160.	562	18	10	1.6	44.2	3346
	563	18	11	2.1	47.2	3724
	564	18	12	2.6	50.2	4102
	565	18	13	3.0	53.0	4480
Fig. 161.	566	18	14	3.5	56.0	4868
	567	18	15	3.9	58.8	5236
	568	18	16	4.4	61.8	5628
	569	18	17	4.9	64.8	6006
	570	18	18	5.3	67.6	6384

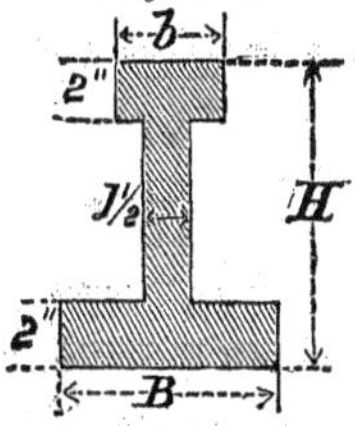

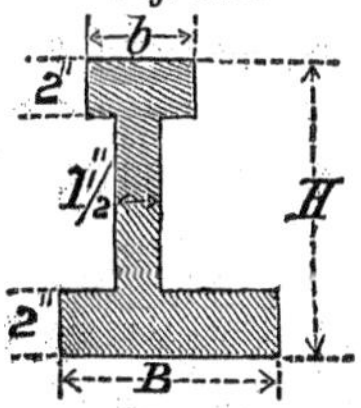

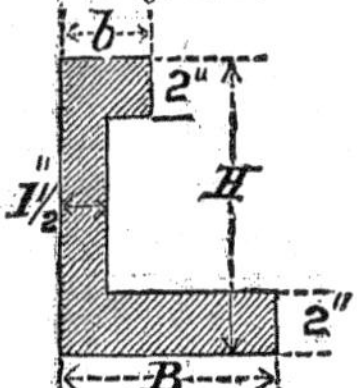

Fig. 162.

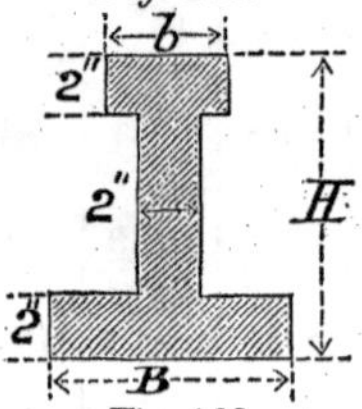

Fig. 163.

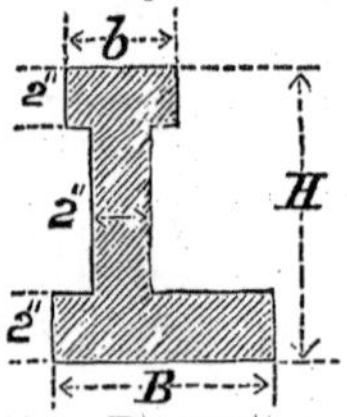

Fig. 164.

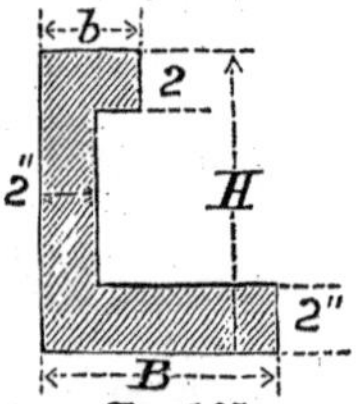

Fig. 165.

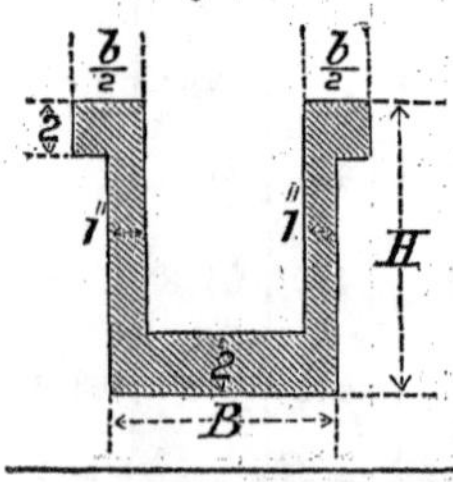

Number of section.	Height H in inches.	Width B of lower flange in inches.	Width b of upper flange in inches.	Sectional area in square inches.	Coefficient K^1.
571	6	9	2.3	26.6	504
572	6	10	2.7	29.4	574
573	6	11	3.0	32.0	630
574	6	12	3.3	34.6	700
575	6	13	3.7	37.4	756
576	6	14	4.0	40.0	826
577	6	15	4.3	42.6	882
578	6	16	4.7	45.5	952
579	6	17	5.0	48.0	1008
580	6	18	5.3	50.6	1064
581	7	9	2.3	28.6	686
582	7	10	2.7	31.4	770
583	7	11	3.0	34.0	854
584	7	12	3.4	36.8	938
585	7	13	3.8	39.6	1036
586	7	14	4.1	42.2	1120
587	7	15	4.5	45.0	1204
588	7	16	4.9	47.8	1288
589	7	17	5.2	50.4	1372
590	7	18	5.6	53.2	1456
591	8	9	2.2	30.4	868
592	8	10	2.6	33.2	980
593	8	11	2.9	35.8	1092
594	8	12	3.3	38.6	1204
595	8	13	3.7	41.4	1302
596	8	14	4.1	44.2	1414
597	8	15	4.5	47.0	1526

Fig. 162.

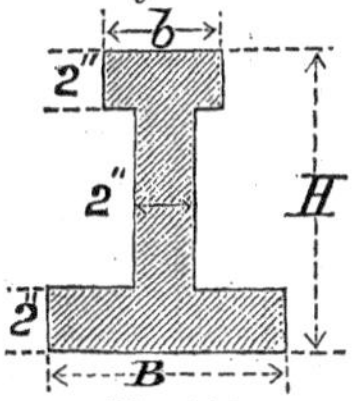

Fig. 163.

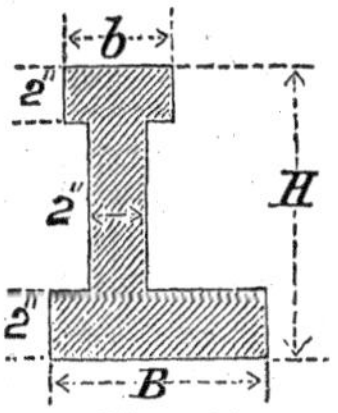

Fig. 164.

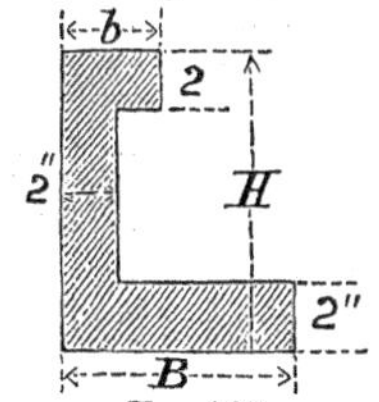

Fig. 165.

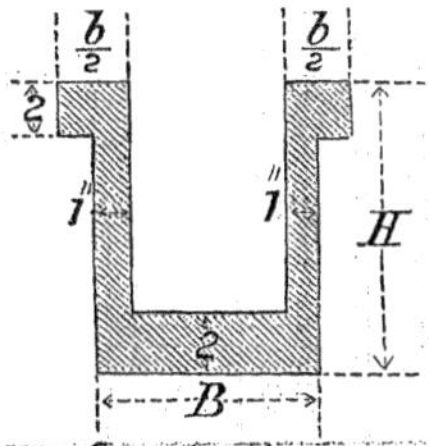

Number of section.	Height H in inches.	Width B of lower flange in inches.	Width b of upper flange in inches.	Sectional area in square inches.	Coefficient K^1.
598	8	16	4.9	49.8	1638
599	8	17	5.3	52.6	1750
600	8	18	5.7	55.4	1848
601	9	9	2.1	32.2	1064
602	9	10	2.5	35.0	1204
603	9	11	2.9	37.8	1330
604	9	12	3.3	40.6	1470
605	9	13	3.7	43.4	1596
606	9	14	4.1	46.2	1736
607	9	15	4.5	49.0	1876
608	9	16	4.9	51.8	2002
609	9	17	5.3	54.6	2142
610	9	18	5.7	57.4	2282
611	10	10	2.4	36.8	1414
612	10	11	2.8	39.6	1582
613	10	12	3.2	42.4	1736
614	10	13	3.6	45.2	1904
615	10	14	4.0	48.0	2058
616	10	15	4.4	50.8	2226
617	10	16	4.8	53.6	2380
618	10	17	5.2	56.4	2595
619	10	18	5.7	59.4	2702
620	11	10	2.2	38.4	1638
621	11	11	2·6	41.2	1820
622	11	12	3.0	44.0	2016
623	11	13	3.5	47.0	2198
624	11	14	3.9	49.8	2380

Fig. 162.

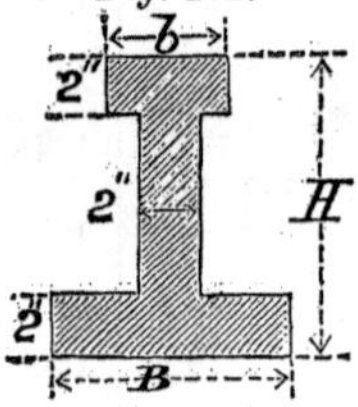

Fig. 163.

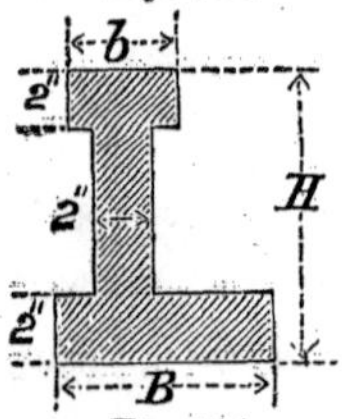

Fig. 164.

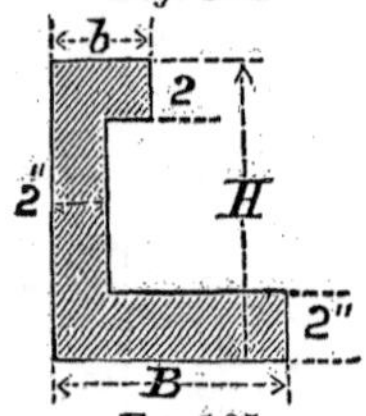

Fig. 165.

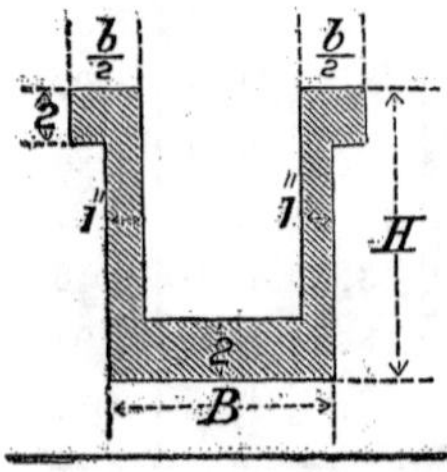

Number of section.	Height H in inches.	Width B of lower flange in inches.	Width b of upper flange in inches.	Sectional area in square inches.	Coefficient K^1.
625	11	15	4.3	52.6	2576
626	11	16	4.7	55.4	2758
627	11	17	5.1	58.2	2954
628	11	18	5.6	61.2	3136
629	12	11	2.4	42.8	2086
630	12	12	2.9	45.8	2296
631	12	13	3.3	48.1	2506
632	12	14	3.7	51.4	2716
633	12	15	4.1	54.2	2940
634	12	16	4.6	57.2	3150
635	12	17	5.0	60.0	3360
636	12	18	5.4	62.8	3570
637	13	11	2.2	44.4	2338
638	13	12	2.7	47.4	2576
639	13	13	3.1	50.2	2814
640	13	14	3.5	53.0	3052
641	13	15	4.0	56.0	3290
642	13	16	4.4	58.8	3528
643	13	17	4.9	61.8	3780
644	13	18	5.3	64.6	4018
645	14	11	2.0	46.0	2604
646	14	12	2.5	49.0	2870
647	14	13	2.9	51.8	3136
648	14	14	3.4	54.8	3402
649	14	15	3.8	57.6	3668
650	14	16	4.2	60.4	3934
651	14	17	4.7	63.4	4208

Fig. 162.

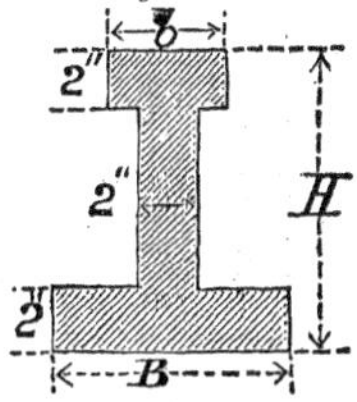

Fig. 163.

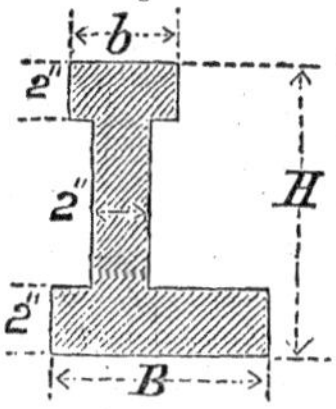

Fig. 164.

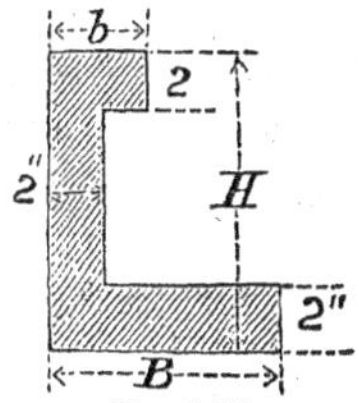

Fig. 165.

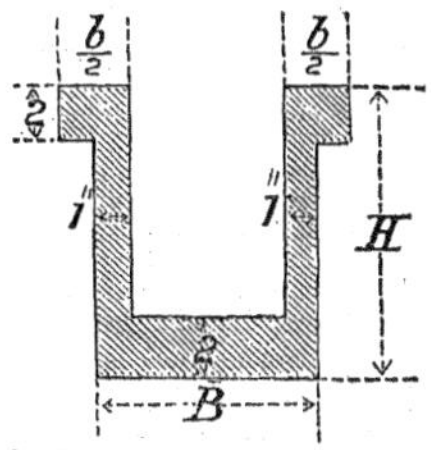

Number of section.	Height *H* in inches.	Width *B* of lower flange in inches.	Width *b* of upper flange in inches.	Sectional area in square inches.	Coefficient *K*1.
652	14	18	5.1	66.2	4452
653	15	12	2.3	50.6	3164
654	15	13	2.7	54.4	3444
655	15	14	3.2	56.4	3738
656	15	15	3.6	59.2	4032
657	15	16	4.1	62.2	4296
658	15	17	4.5	65.0	4606
659	15	18	4.9	67.8	4900
660	16	13	2.5	55.0	3742
661	16	14	3.0	58.0	4074
662	16	15	3.4	60.8	4396
663	16	16	3.9	63.8	4718
664	16	17	4.3	66.6	5026
665	16	18	4.8	69.6	5348
666	17	13	2.3	56.6	4060
667	17	14	2.8	59.6	4410
668	17	15	3.2	62.4	4760
669	17	16	3.7	65.4	5110
670	17	17	4.1	68.2	5460
671	17	18	4.6	71.2	5810
672	18	13	2.1	58.2	4382
673	18	14	2.5	61.0	4746
674	18	15	3.0	64.0	5124
675	18	16	3.4	66.8	5502
676	18	17	3.9	69.8	5080
677	18	18	4.4	72.8	6258

STRENGTH OF WOODEN BEAMS.

Capacity W in lbs. of American white and yellow pine beams, joists, &c., from 1″ x 1″ to 15 x 15 in.

The modulus of rupture is taken at $\frac{10000}{8} = 1250$ lbs., or 8 times safety.

K' = tabulated coefficient, to be divided by
l = distance between supports in inches, or length of beams in inches from support to free end of beam.

Thickness in inches.	Coefficient						
	Height in						
	1	2	3	4	5	6	7
1	1666	6666	15000	26666	41666	60000	81666
1½	2500	10000	22500	39999	62499	90000	122499
2	3333	13333	30000	53333	83333	120000	163333
2½	4166	16666	37500	66666	104166	150000	204166
3	5000	19999	45000	80000	124999	180000	244999
3½	5833	23333	52700	93333	145833	210000	285833
4	6666	26666	60000	106666	166666	240000	326666
4½	7499	29999	67500	119999	187499	270000	367499
5	8333	33333	75000	133333	208333	300000	408333
5½	9166	36666	82500	146666	229166	330000	449166
6	10000	39999	90000	159999	249999	360000	489999
6½	10833	43333	97500	173333	270833	390000	530833
7	11666	46666	105000	186666	291666	420000	571666
7½	12500	49999	112600	199999	312499	450000	612499
8	13333	53333	120000	213333	333333	480000	653333
8½	14166	56666	127500	226666	354166	510000	694166
9	14998	59999	135000	239999	374999	540000	734999
9½	15831	63333	142500	253333	395833	570000	775833
10	16666	66666	150000	266666	416666	600000	816666
10½	17500	69999	157500	279999	437499	630000	857599
11	18333	73333	165000	293333	458333	660000	898533
11½	19166	76666	172500	306666	479166	690000	939366
12	20000	79999	180000	319999	499999	720000	979999
12½	20833	83333	187500	333333	520833	750000	1020833
13	21666	86666	195000	346666	541666	780000	1061666
13½	22500	89999	202500	359999	562499	810000	1102499
14	23333	93333	210000	373333	583333	840000	1143333
14½	24166	96666	217500	386666	604166	870900	1184166
15	25000	99999	225000	399999	624999	900000	1224999

BEAMS SUPPORTED AT THE ENDS.

Load equally distributed, $W = \frac{K'}{l}$ or $K' = lW$. 1

Load concentrated at centre, $W = \frac{K'}{2l}$ or $K' = 2lW$. 2

BEAMS FIXED AT ONE END.

Load equally distributed, $W = \frac{K'}{4l}$ or $K' = 4lW$. 3

Load concentrated at free end, $W = \frac{K'}{8l}$ or $K' = 8lW$. 4

K'.

inches.

8	9	10	11	12	13	14	15
106666	135000	166666	201757	240000	281000	320000	375000
159999	202500	249999	302636	360000	422499	489999	562500
213333	270000	333333	403515	480000	563333	653333	750000
266666	337500	416666	504393	600000	704166	816666	937500
319999	405000	499999	605272	720000	844999	979999	1125000
373333	472500	583333	706151	840000	985833	1143333	1312500
426666	540000	666666	807030	960000	1126666	1306666	1500000
479999	607500	749999	907908	1080000	1267499	1469999	1687500
533333	675000	833333	1008787	1200000	1408333	1633333	1875000
586666	742500	916666	1109666	1320000	1549166	1796666	2062500
639999	810000	999999	1210545	1440000	1689999	1959999	2250000
693333	877500	1083333	1311423	1560000	1830833	2123333	2437500
746666	945000	1166666	1412202	1680000	1971666	2286666	2625000
799999	1012500	1249999	1513181	1800000	2112499	2449999	2812500
853333	1080000	1333333	1614060	1920000	2253333	2613333	3000000
906666	1147500	1416666	1714938	2040000	2394166	2776666	3187500
959999	1215000	1499999	1815817	2160000	2534999	2939999	3375000
1013333	1282500	1583333	1916696	2280000	2675833	3103333	3562500
1066666	1350000	1666666	2017575	2400000	2816666	3266666	3750000
1119999	1417500	1749999	2118453	2520000	2957499	3429999	3937500
1173333	1485000	1833333	2219332	2640000	3098333	3593333	4125000
1226666	1552500	1916666	2320211	2760000	3239166	3756666	4312500
1279999	1620000	1999999	2421090	2880000	3379999	3919999	4500000
1333333	1687500	2083333	2521968	3000000	3520833	4083333	4687500
1386666	1755000	2166666	2622847	3120000	3661666	4246666	4875000
1439999	1822500	2249999	2723726	3240000	3802499	4409999	5062500
1493333	1890000	2333333	2824605	3360000	3943333	4573333	5250000
1546666	1957500	2416666	2925483	3480000	4084166	4736666	5437500
1599999	2025000	2499999	3026362	3600000	4224999	4899999	5625000

PRESSURE ON SUPPORTS.

Reaction of Supports.

For a continuous beam, horizontal or inclined. Load $W_{/}$ equally distributed, and supports equal distance apart. Applicable to trussed beams, rafters, or beams supported by three or more supports.

Reference. (*Fig.* 166.)

$W_{/}$ = Weight of load per unit of length in lbs.
L = Distance between supports in units of length.
P, P_1, P_2 = Pressure on supports in lbs., counting from end support to center of beam.
M, M_1, M_2 = Moments of rupture over supports.
m, m_1, m_2 = Moments of rupture between supports.
l, l_1, l_2 = The distance from a support to section where moments m, m_1, m_2 occur.

By this table the pressure upon any support, from 3 to 9 in number, can be ascertained; also the moments of rupture. The table is used in calculating the strains in roof trusses, &c.

Fig. 166.

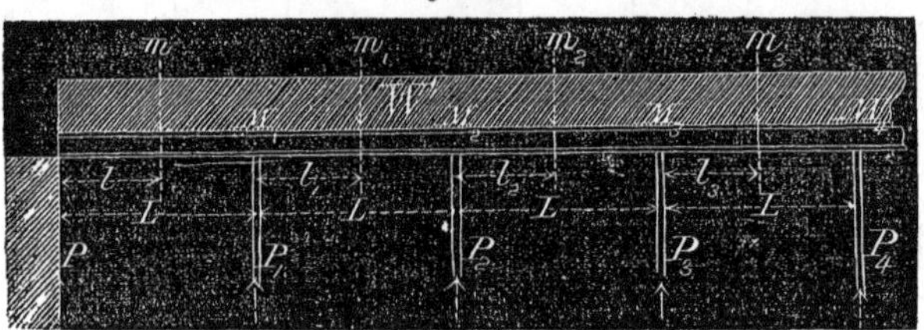

Reactions or pressure.	Number of Supports.				
	3	4	5	7	9
P	0.375 $W_{/}L$	0.4 $W_{/}L$	0.3929 $W_{/}L$	0.3942 $W_{/}L$	0.3943 $W_{/}L$
P_1	1.25 $W_{/}L$	1.1 $W_{/}L$	1.1429 $W_{/}L$	1.1346 $W_{/}L$	1.1340 $W_{/}L$
P_2	....	....	0.9286 $W_{/}L$	0.9615 $W_{/}L$	0.9629 $W_{/}L$
P_3	....	....	....	1.0192 $W_{/}L$	1.0103 $W_{/}L$
P_4	....	....	....	....	0.9948 $W_{/}L$
M_1	0.125 $W_{/}L^2$	0.1 $W_{/}L^2$	0.1071 $W_{/}L^2$	0.1058 $W_{/}L^2$	0.1057 $W_{/}L^2$
M_2	....	....	0.0714 $W_{/}L^2$	0.0769 $W_{/}L^2$	0.0773 $W_{/}L^2$
M_3	....	....	....	0.0865 $W_{/}L^2$	0.0850 $W_{/}L^2$
M_4	....	....	....	....	0.0824 $W_{/}L^2$

Reactions or pressure.	Number of Supports.				
	3	4	5	7	9
m	$0.0703\,W_1L^2$	$0.08\ \ W_1L^2$	$0.0772\,W_1L^2$	$0.0777\,W_1L^2$	$0.0777\,W_1L^2$
m_1	....	$0.025\ \ W_1L^2$	$0.0364\,W_1L^2$	$0.0340\,W_1L^2$	$0.0339\,W_1L^2$
m_2	....	....	....	$0.0434\,W_1L^2$	$0.0438\,W_1L^2$
m_3	....	....	....	....	$0.0412\,W_1L^2$
l	$0.375\ L$	$0.4\ L$	$0.3928\ L$	$0.3942\ L$	$0.3943\ L$
l_1	....	$0.5\ L$	$0.535\ \ L$	$0.5288\ L$	$0.5283\ L$
l_2	....	....	....	$0.4903\ L$	$0.4922\ L$
l_3	....	....	....	....	$0.5025\ L$

Reference. (*Figs.* 167, 168, and 169.)

$W,\ W_1,\ W_2$ = Load in lbs.

$l,\ l_1,\ l_2$ = Dimensions in units of length.

$P,\ P_1,\ P_2$ = Pressure on supports in lbs.

Fig. **167.**

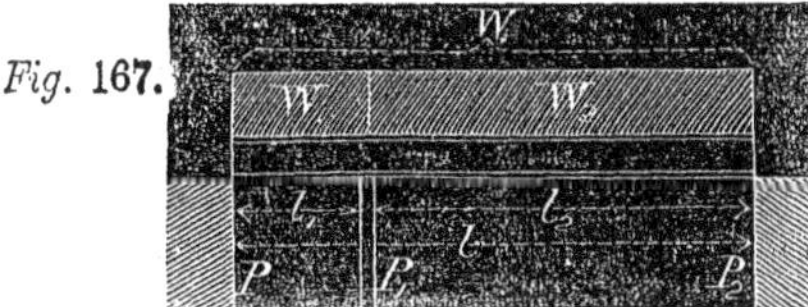

Three supports, unequal distances apart.

Load equally distributed:

$l_1 < l_2$;

$$W_1 = \frac{l_1}{l} W \qquad P = \tfrac{3}{8} W_1 = \tfrac{3}{8} \frac{l_1}{l} W$$

$$P_1 = \tfrac{5}{8}(W_1 + W_2) = \tfrac{5}{8} W$$

$$W_2 = \frac{l_2}{l} W \qquad P_2 = \tfrac{3}{8} W_2 = \tfrac{3}{8} \frac{l_2}{l} W$$

Fig. 168.

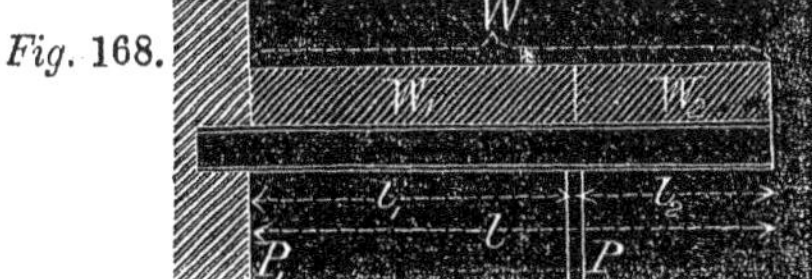

One support, and fixed at one end.

Load equally distributed:

$l_1 > l_2$

$$W_1 = \frac{l_1}{l} W \qquad P = \tfrac{1}{2} \frac{Wl}{l_1}$$

$$W_2 = \frac{l_2}{l} W \qquad P_1 = W - P = \left(1 - \tfrac{1}{2} \frac{l}{l_2}\right) W$$

Fig. 169.

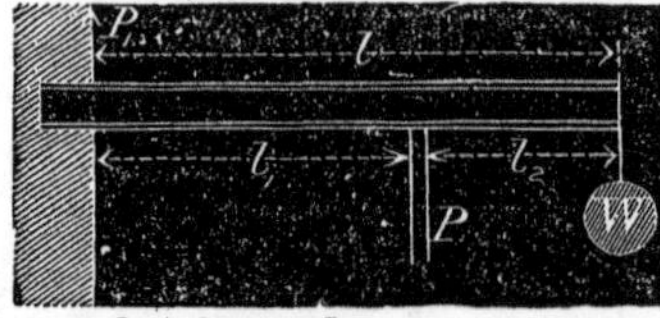

One support, and fixed at one end.

Load concentrated at free end:

$$P = \frac{l}{l_1} W$$

$$P_1 = P - W = \left(\frac{l}{l_1} - 1\right) W = \frac{l_2}{l_1} W$$

COMPRESSIVE STRAIN AND PRESSURE ON SUPPORTS.

Sloping Beams, Rafters, &c.

Load W equally distributed.

For the cross-breaking strain, the rafter, &c., is to be treated as a horizontal beam of the length l. (See *Compound Strains in Beam, &c.*)

Reference.

C = Compression in direction of beam.
H = Horizontal strain acting on support.
V = Pressure on supports.

Lower end supported vertically and horizontally; upper end resting on inclined support:

Fig. 170.

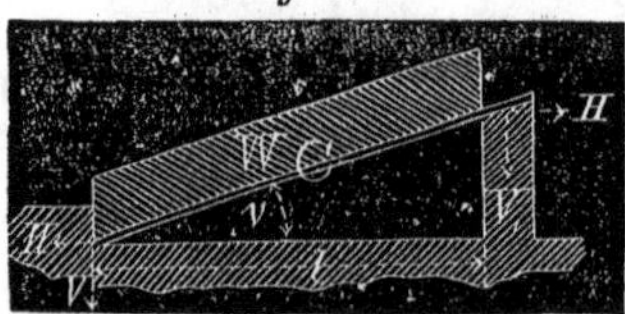

$$C = \frac{W}{2} \sin.v \qquad V = W - V_1 = W\left(1 - \tfrac{1}{2} (\cos.v)^2\right)$$

$$H = \frac{W}{2} \sin.v \cos.v \qquad V_1 = \frac{W}{2} (\cos.v)^2$$

Upper end fixed; lower end supported horizontally:

Fig. 171.

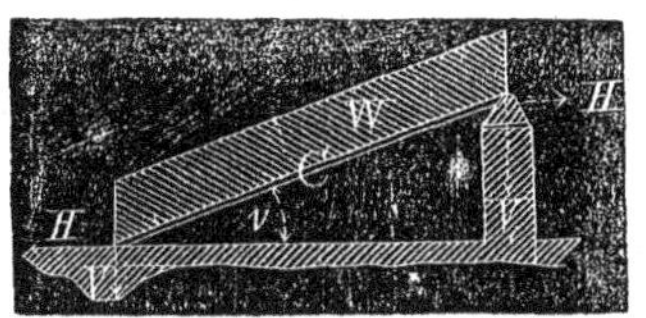

$$C = 0$$
$$H = 0$$
$$V = V_1 = \frac{W}{2}$$

Upper end resting against a vertical surface; lower end supported vertically and horizontally:

Fig. 172.

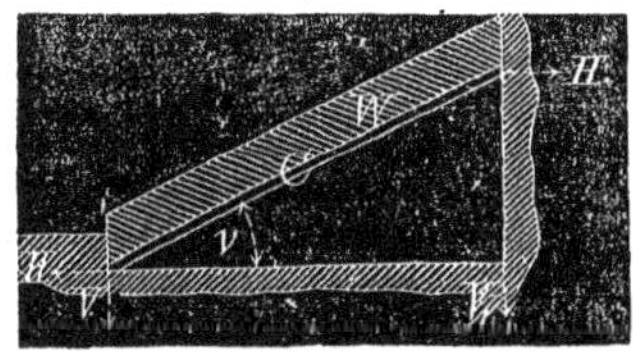

$$C = \frac{W}{2 \sin . v}$$
$$H = \frac{W}{2} \text{ cotg } v$$
$$V = W$$
$$V_1 = 0$$

RESISTANCE TO CRUSHING.

Strength of Columns, Pillars, and Struts.

Reference.

A = Area of cross-section in inches.
C = Coefficient, depending on the material.
I = Least moment of inertia of cross-section.
W = Capacity of column, pillar, or strut in lbs.
a = Coefficient, depending on the material in respect to flexure.
c = Coefficient, depending on the material.
h = The least dimension across the section in inches.
k = Factor of safety.
l = Length of column, &c., in inches.
r = Least radius of gyration.

To find the square of the radius of gyration (r^2) of a plane about a given axis, divide the least moment of inertia by the sectional area of the plane; that is, $r^2 = \frac{I}{A}$.

Values of—	For Malleable Iron.	For Cast Iron.	For Dry Timber.
$C =$	36,000 lbs.	80,000 lbs.	7,200 lbs.
$c =$	36,000 "	3,200 "	3,000 "
$a =$	0.000333	0.0025	0.004

The factor of safety k should be, for wrought iron = 6; for cast iron = 8; for timber = 10. This applies to moving loads.

Case 1.

Rounded or hinged at both ends, as per—

Fig. 173.

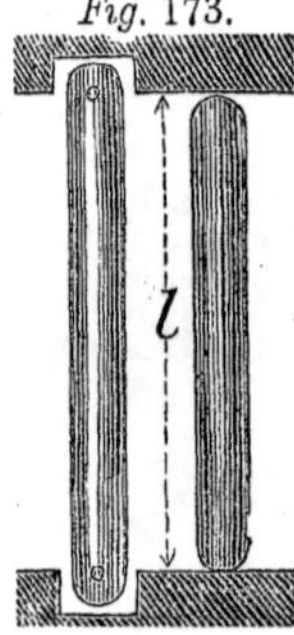

For square, rectangular, or circular cross-section:

$$W = \frac{1}{k} \frac{CA}{1 + 4a \frac{l^2}{h^2}}$$

For any other cross-section:

$$W = \frac{1}{k} \cdot \frac{CA}{1 + \frac{4l^2}{cr^2}}$$

Case 2.

Fixed, or having a flat base at one end, and rounded or hinged at the other, as per—

Fig. 174.

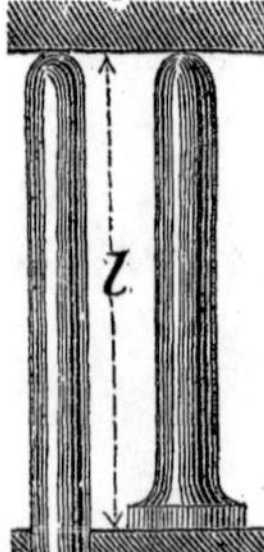

For square, rectangular, or circular cross-section:

$$W = \frac{1}{k} \frac{CA}{1 + 2a \frac{l^2}{h^2}}$$

For any other cross-section:

$$W = \frac{1}{k} \frac{CA}{1 + \frac{16.l^2}{9.c.r^2}}$$

Case 3.

Fixed, or having flat bases at both ends, as per—

Fig. 175.

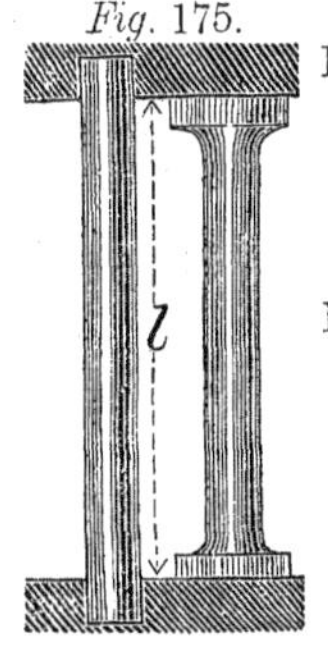

For square, rectangular, or circular cross-section:

$$W = \frac{1}{k} \frac{CA}{1 + a\frac{l^2}{h^2}}$$

For any other cross-section:

$$W = \frac{1}{k} \frac{CA}{1 + \frac{l^2}{c.r^2}}$$

EXAMPLES.

Case 1.

Rounded at both ends:

What is the capacity of a *wrought-iron* strut of the annexed figure and dimensions?

$l = 10$ feet $= 120$ inches.

$A = 4.68$ inches.

Fig. 176.

$$I = \frac{0.9 \times 3.5^3 + 5.1 \times 0.3^3}{12} = 3.227$$

$$r^2 = \frac{3.227}{4.68} = 0.689$$

$$W = \tfrac{1}{4} \frac{36000 \times 4.68}{1 + \frac{4 \times 120^2}{36000 \times 0.689}} = \tfrac{1}{4} \frac{168480}{1 + \frac{57600}{24804}} =$$

$$\tfrac{1}{4} \frac{168480}{3.322} = 12{,}679 \text{ lbs.}$$

The same as above, in Case 3, fixed at both ends:

$$W = \tfrac{1}{4} \frac{36000 \times 4.69}{1 + \frac{120^2}{36000 \times 0.689}} = \tfrac{1}{4} \frac{168480}{1 + \frac{14400}{24804}} =$$

$$\tfrac{1}{4} \frac{168480}{1.58} = 26{,}677 \text{ lbs.}$$

For the annexed figure and dimensions; otherwise, same as above:

$A = 7$ inches.

Case 1.

Rounded at both ends:

Fig. 177.

$$I = \frac{1 \times 4^3 + 3 \times 1^3}{12} = 5.6$$

$$r^2 = \frac{5.6}{7} = 0.8$$

$$W = \tfrac{1}{4}\,\frac{36000 \times 7}{1 + \dfrac{4 \times 120^2}{36000 \times 0.8}} = \tfrac{1}{4}\,\frac{252000}{3} = 21{,}000 \text{ lbs.}$$

Same as above, in Case 3, fixed at both ends:

$$W = \tfrac{1}{4}\,\frac{36000 \times 7}{1 + \dfrac{120^2}{36000 \times 0.8}} = \tfrac{1}{4}\,\frac{252000}{1.5} = 42{,}000 \text{ lbs.}$$

Case 3.

Fixed ends:

What is the capacity of a *cast-iron* pillar of the annexed figure and dimensions?

$l = 10$ feet $= 120$ inches.
$A = 11$ inches.

Fig. 178.

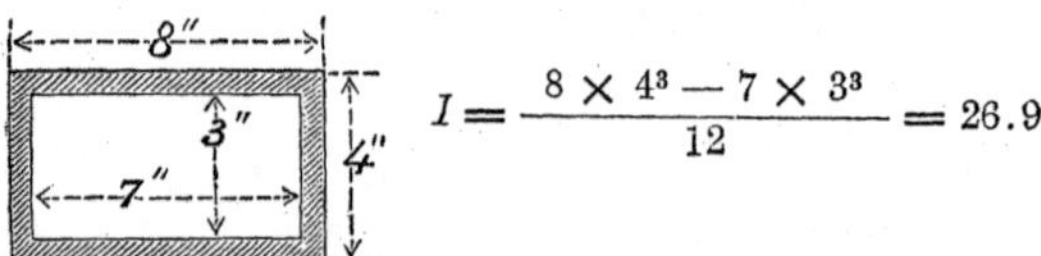

$$I = \frac{8 \times 4^3 - 7 \times 3^3}{12} = 26.9$$

$$W = \tfrac{1}{8}\,\frac{80000 \times 11}{1 + 0.0025\,\dfrac{120^2}{4^2}} = \tfrac{1}{8}\,\frac{880000}{3.25} = 33{,}846 \text{ lbs.}$$

For the annexed figure and dimensions; otherwise, same as above.

Fig. 179.

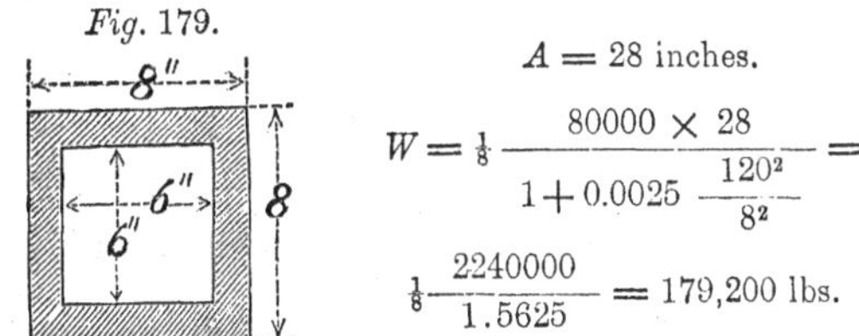

$A = 28$ inches.

$$W = \frac{1}{8} \frac{80000 \times 28}{1 + 0.0025 \frac{120^2}{8^2}} =$$

$$\frac{1}{8} \frac{2240000}{1.5625} = 179{,}200 \text{ lbs.}$$

For the annexed figure and dimensions; otherwise, same as above.

Fig. 180.

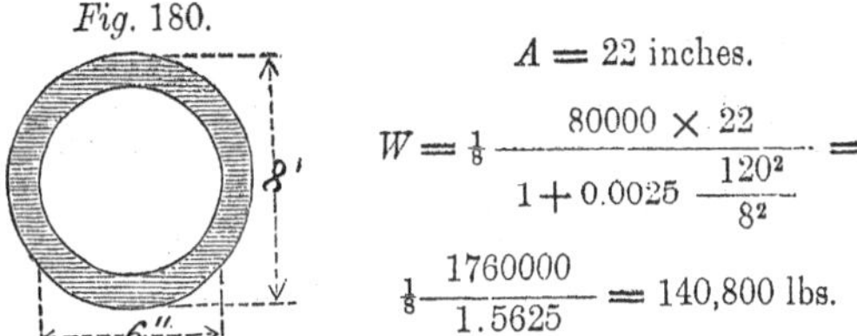

$A = 22$ inches.

$$W = \frac{1}{8} \frac{80000 \times 22}{1 + 0.0025 \frac{120^2}{8^2}} =$$

$$\frac{1}{8} \frac{1760000}{1.5625} = 140{,}800 \text{ lbs.}$$

To find the capacity of a Column, Pillar, or Strut of any cross-section by the following Table:

Find how many times the least dimension h across the section is contained in the length l of column, &c.—that is, $\frac{l}{h}$—then multiply the corresponding number on the same horizontal line, under K'', by the sectional area of cross-section. This gives the capacity in tons of 2,000 lbs.

Let l = Length of column, &c.

h = Least dimension of cross-section.

K'' = Capacity in tons of one square inch of cross-section, to be multiplied by sectional area of desired cross-section.

Various sections for which this table is applicable:

Fig. 181. *Fig.* 182.

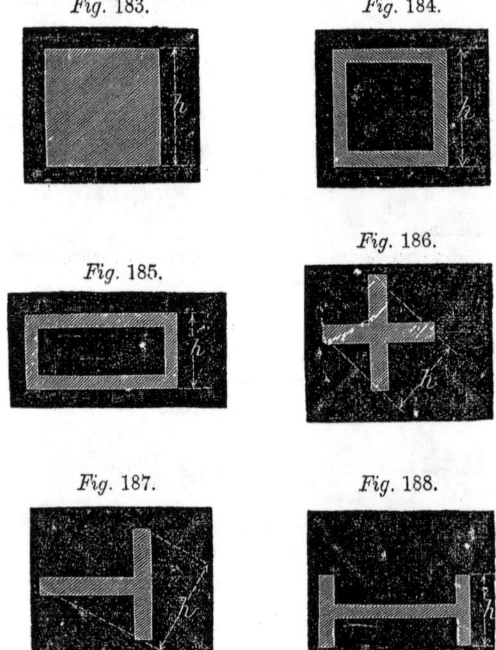

Fig. 183. *Fig.* 184. *Fig.* 185. *Fig.* 186. *Fig.* 187. *Fig.* 188.

[NOTE.—This table is strictly correct, only for columns, &c., with circular or rectangular cross-section. As the error is small, it may be used for any cross-section.]

Example explanatory of the following table.

What is the capacity of a cast-iron column 10 feet = 120 inches long, fixed at both ends, and of the annexed cross-section and dimensions?

Fig. 189.

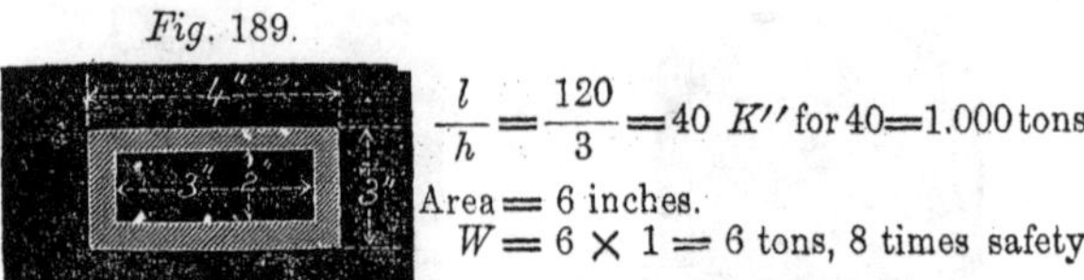

$\frac{l}{h} = \frac{120}{3} = 40$ K'' for 40=1.000 tons.

Area = 6 inches.

$W = 6 \times 1 = 6$ tons, 8 times safety.

Column, &c., fixed at both ends.

Cast Iron—eight times safety.						Wrought Iron—six times safety.					
$\frac{l}{h}$	K''	$\frac{l}{h}$	K''	$\frac{l}{h}$	K''	$\frac{l}{h}$	K''	$\frac{l}{h}$	K''	$\frac{l}{h}$	K''
	Tons.		Tons.		Tons.		Tons.		Tons.		Tons.
1	4.987	25	1.951	49	0.714	1	2.999	25	2.487	49	1.674
2	4.950	26	1.858	50	0.689	2	2.996	26	2.452	50	1.644
3	4.890	27	1.771	51	0.666	3	2.991	27	2.418	51	1.615
4	4.807	28	1.689	52	0.644	4	2.984	28	2.383	52	1.585
5	4.705	29	1.611	53	0.623	5	2.975	29	2.348	53	1.557
6	4.587	30	1.538	54	0.603	6	2.964	30	2.313	54	1.529
7	4.450	31	1.469	55	0.584	7	2.953	31	2.277	55	1.501
8	4.310	32	1.404	56	0.565	8	2.938	32	2.242	56	1.474
9	4.158	33	1.343	57	0.548	9	2.921	33	2.206	57	1.448
10	4.000	34	1.285	58	0.531	10	2.905	34	2.172	58	1.422
11	3.838	35	1.230	59	0.515	11	2.885	35	2.136	59	1 396
12	3.676	36	1.179	60	0.500	12	2.863	36	2.101	60	1.371
13	3.514	37	1.130	61	0.485	13	2.841	37	2.067	61	1.347
14	3.355	38	1.084	62	0.471	14	2.817	38	2.032	62	1.323
15	3.200	39	1.041	63	0.457	15	2.792	39	1.998	63	1.299
16	3.048	40	1.000	64	0.445	16	2.766	40	1.963	64	1.276
17	2.902	41	0.961	65	0.432	17	2.738	41	1.930	65	1.253
18	2.762	42	0.924	66	0.420	18	2.711	42	1.896	66	1.228
19	2.628	43	0.889	67	0.409	19	2.680	43	1.863	67	1.209
20	2.500	44	0.856	68	0.398	20	2.650	44	1.831	68	1.187
21	2.378	45	0.824	69	0.387	21	2.619	45	1.798	69	1.167
22	2.262	46	0.794	70	0.377	22	2.586	46	1.767	70	1.146
23	2.152	47	0.766	71	0.367	23	2.554	47	1.735	71	1.126
24	2.049	48	0.739	72	0.358	24	2.520	48	1.704	72	1.107

Strength of Columns, Pillars, or Struts, of seasoned wood, round or square section.

Fixed at both ends. All dimensions in inches.

Find how many times the least dimension across the section is contained in the length or height of column, &c.; that is, $\frac{H}{D}$; then multiply the corresponding figures on the same horizontal line under K'' by the sectional area of cross-section. This gives the capacity of column, &c., in tons of 2,000 lbs., 10 times safety.

Reference.

$H =$ Length of column, &c.

$D =$ Least dimension of cross-section.

$K'' =$ Capacity in tons of one square inch of cross-section, to be multiplied by sectional area of desired cross-section.

The coefficient C for white and yellow pine in the following table is taken at $\frac{6000}{10} = 600$ lbs. for safety:

For oak at $\frac{8000}{10} = 800$ lbs. per square inch for safety.

EXAMPLE.—What is the capacity of a pillar of oak, section 4×6 inches, length $= 12$ feet $= 144$ inches?

$$\frac{H}{D} = \frac{144}{4} = 36,\ K'' \text{ for } 36 = 0.064 \times 4 \times 6 = 1.536 \text{ tons.}$$

Capacity K'' of one square inch in tons of 2,000 lbs.

White and Yellow Pine.				Oak.			
$\frac{H}{D} =$	K''	$\frac{H}{D} =$	K''	$\frac{H}{D} =$	K''	$\frac{H}{D} =$	K''
1	0.299	26	0.081	1	0.399	26	0.108
2	0.295	27	0.076	2	0.394	27	0.102
3	0.289	28	0.072	3	0.386	28	0.096
4	0.282	29	0.068	4	0.376	29	0.091
5	0.272	30	0.065	5	0.363	30	0.086
6	0.262	31	0.061	6	0.349	31	0.082
7	0.251	32	0.058	7	0.334	32	0.078
8	0 239	33	0.056	8	0.319	33	0.074
9	0.226	34	0.053	9	0.302	34	0.071
10	0.214	35	0.050	10	0.285	35	0.067
11	0.202	36	0.048	11	0 269	36	0.064
12	0.190	37	0.046	12	0.254	37	0.061
13	0.179	38	0.044	13	0.238	38	0.059
14	0.168	39	0.042	14	0.224	39	0.056
15	0.158	40	0.040	15	0.210	40	0.054
16	0.148	41	0.038	16	0.197	41	0.051
17	0.139	42	0.037	17	0.185	42	0.049
18	0.130	43	0.035	18	0.174	43	0.047
19	0.123	44	0.034	19	0.163	44	0.045
20	0.115	45	0.033	20	0.154	45	0.044
21	0.108	46	0.031	21	0.144	46	0.042
22	0.102	47	0.030	22	0.136	47	0.040
23	0.096	48	0.029	23	0.128	48	0.039
24	0.090	49	0.028	24	0.121	49	0.037
25	0.085	50	0.027	25	0.114	50	0.036

PARALLELOGRAM OF FORCES.

Composition and Resolution of Forces.

Reference.

A, B, C = Forces, or strains, acting on a single point.
v, v', = angles.

Fig. 190.

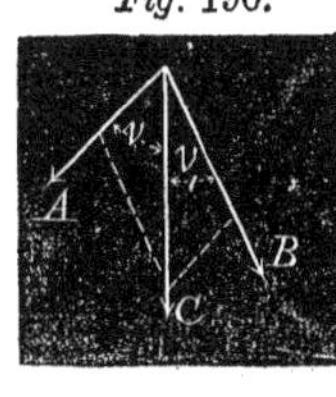

$$A = \frac{C \sin. v_{\prime}}{\sin. (v + v_{\prime})},$$

$$B = \frac{C \sin. v}{\sin. (v + v_{\prime})}, \text{ when } v = v_{\prime},\ A = B = \frac{C}{2} \sec. v;$$

when $v + v_{\prime} < 90°$ $\quad C = \sqrt{A^2 + B^2 + \left(2\,A\,B \cos. (v + v_{\prime})\right)}$

when $v + v_{\prime} > 90°$ $\quad C = \sqrt{A^2 + B^2 - \left[2\,A\,B \cos. \left(180° - (v + v_{\prime})\right)\right]}$

Fig. 191.

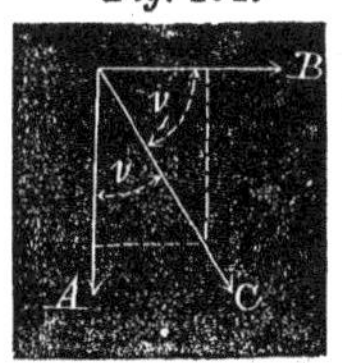

$$v + v_{\prime} = 90°$$
$$A = C \cos. v$$
$$B = C \sin. v = C \cos. v_{\prime}$$
$$C = \sqrt{A^2 + B^2}$$

Fig. 192.

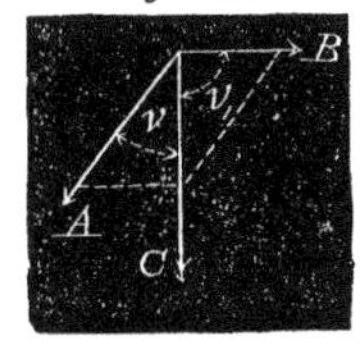

$$v_{\prime} = 90°$$
$$A = \frac{C}{\cos. v}$$
$$B = C \text{ tang. } v$$
$$C = \sqrt{A^2 - B^2}$$

STRAINS IN FRAMES.

Reference.

C = Compressive strain in units of weight.
T = Tensile " "
V = Vertical " "
H = Horizontal " "
W = Load in units of weight.
l = Dimensions in units of length.
v = Angle between horizontal and inclined member.

For cross-breaking strain, see "Resistance to cross-breaking."

Fig. 193.

$$C = \frac{W}{2 \sin. v}$$

$$C_{,} = \frac{W}{2} \text{cotg. } v = H$$

Fig. 194.

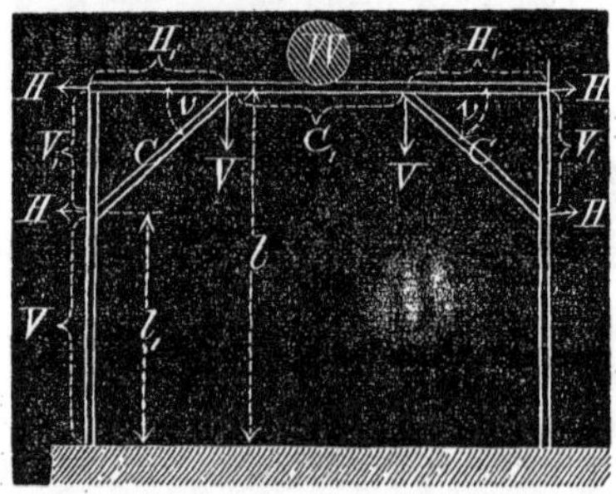

$$C = \tfrac{11}{16} \frac{W}{\sin. v}$$

$C_{,} = H = \tfrac{11}{16}\ W \text{cotg.} v$ = cross-breaking strain at H.

$H_{,} = \frac{l_{,}}{l}\ H = \tfrac{11}{16} . \frac{l_{,}}{l}\ W \text{cotg. } v$ = tension in $H_{,}$.

$H - H_{,} = \tfrac{11}{16} . \left(\frac{l - l_{,}}{l}\right) W \text{cotg. } v$ = compression in $C_{,}$

$V = \tfrac{11}{16}\ W.$

$V_{,} = \tfrac{3}{16}\ W.$

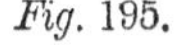

Fig. 195.

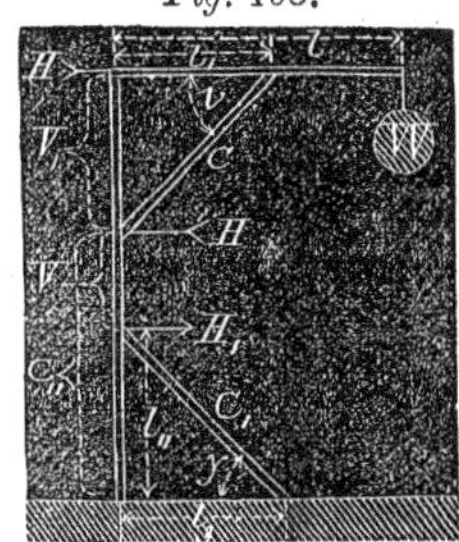

$$C = \frac{l W}{l_{,} \sin. v} = \text{compression.}$$

$$C_{,} = \frac{H_{,}}{\cos. y} = \frac{W.l}{l_{,,}.\cos. y} = \text{compression.}$$

$$C_{,,} = W.$$

$$H = W.l.$$

$$H_{,} = \frac{W.l}{l_{,,}}$$

$$V = H_{,} \text{ tang. } y = \frac{W.l}{l_{,,}} \text{ tang. } y$$

When $l > l_3$ the portion $l_{,,}$ is in tension $= V - W =$

$$W\left(\frac{l}{l_{,,}} \text{ tang. } y - 1\right)$$

When $l < l_3$ the portion $l_{,,}$ is in compression $= W - V =$

$$W\left(1 - \frac{l}{l_{,,}} \text{ tang. } y\right)$$

$$V_{,} = \frac{l - l_{,}}{l_{,}} . W = \text{tension.}$$

Fig. 196.

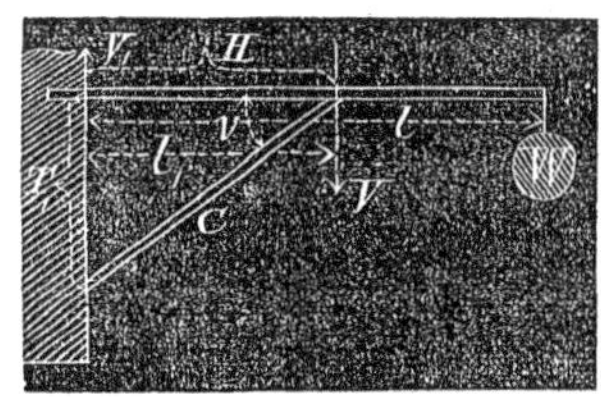

Fig. 197.

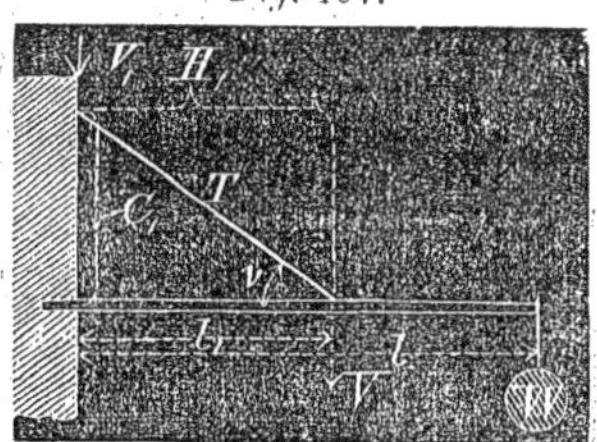

Ends of beams built into wall or fixed:

$$V = \frac{l}{l_{,}} W$$

$$V_{,} = V - W = \left(\frac{l - l_{,}}{l_{,}}\right) W_{,} = T_{,} \text{ (tension)} = C_{,} \text{ (compression.)}$$

$$C = \left(\frac{3l - l_{\prime}}{2l_{\prime}}\right) \frac{W}{\text{sin. } v} = \text{(compression)} = T \text{ (tension.)}$$

$$H = \left(\frac{3l - l_{\prime}}{2l_{\prime}}\right) W \text{ cotg. } v = \text{(tension)} = H_{\prime} \text{ (compression.)}$$

Ends of beams *not* built into wall or fixed:

$$V = \frac{l}{l_{\prime}} W$$

$$V_{\prime} = V - W = \left(\frac{l - l_{\prime}}{l_{\prime}}\right) W = C_{\prime} \text{ (compression)} = T_{\prime} \text{ (tension.)}$$

$$C = \frac{V}{\text{sin. } v} = \frac{l W}{l_{\prime} \text{ sin. } v} = T \text{ (tension.)}$$

$$H = V \text{ cotg. } v = \frac{l}{l_{\prime}} W \text{ cotg. } v = \text{(tension)} = H_{\prime} \text{ (compression.)}$$

STRAINS IN BOOM DERRICKS.

Reference.

C = Compression in boom.
$C_{\prime}$ = Compression in mast.
T = Tension in tackling.
$T_{\prime}$ = Tension in guy.
t = Tension in runner from mast head to weight.
$t_{\prime}$ = Tension in runner from boom head to weight.
W = Weight or load.
H = Horizontal strain.
V = Vertical strain.
v, v_1, v_2 = Angles. (See Figure.)

Fig. 198.

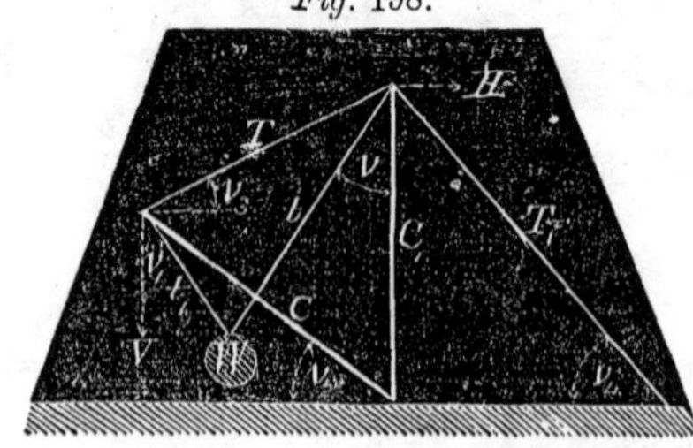

$$t = \frac{W \sin. v_1}{\sin. (v + v_1)} \qquad t_{\prime} = \frac{W \sin. v}{\sin. (v + v_1)}$$

$V = t_{\prime} \text{ cosin. } v_1$ $\qquad H = V \text{ cotg. } v_3$

$C = V \text{ cosec. } v_2$ $\qquad C_{\prime} = W$

$T = V \text{ cosec. } v_3$ $\qquad T_{\prime} = V \text{ cotg. } v_3 \text{ sec. } v_4$

STRAINS IN TRUSSES.

Load equally distributed.

Reference.

W = Load equally distributed in lbs.
l = Distance between abutments.
v = Angle between horizontal and diagonal.
C = Compression in lbs., (denoted by thick lines.)
T = Tension in lbs., (denoted by thin lines.)

2 Bays $= \frac{l}{2}$

Fig. 199.

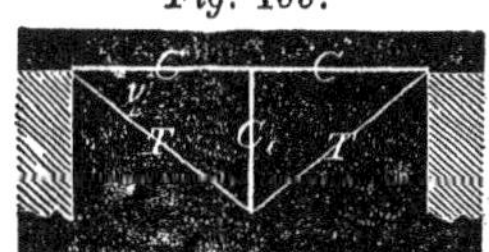

$$C = \tfrac{5}{16} W \text{ cotg. } v$$
$$C_1 = \tfrac{5}{8} W$$
$$T = \tfrac{5}{16} \frac{W}{\sin. v}$$

3 Bays $= \frac{l}{3}$

Fig. 200.

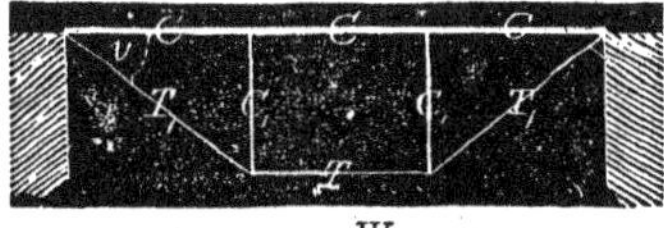

$$C = T = \frac{W}{3} \text{ cotg. } v$$

$$C_1 = \frac{W}{3}$$

$$T_1 = \tfrac{1}{3} \frac{W}{\sin. v}$$

4 Bays $= \frac{l}{4}$

Fig. 201.

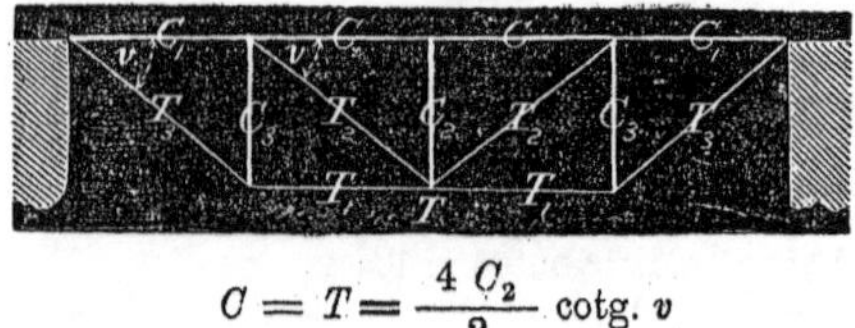

$$C = T = \frac{4\,C_2}{2} \text{ cotg. } v$$

$$C_1 = T_1$$

$$C_2 = \frac{W}{4}$$

$$C_3 = \frac{3C_2}{2}$$

$$T_1 = \frac{3C_2}{2} \text{ cotg. } v$$

$$T_2 = \frac{C_2}{2} \text{ cosec. } v$$

$$T_3 = 3T_2$$

5 Bays $= \frac{l}{5}$

Fig. 202.

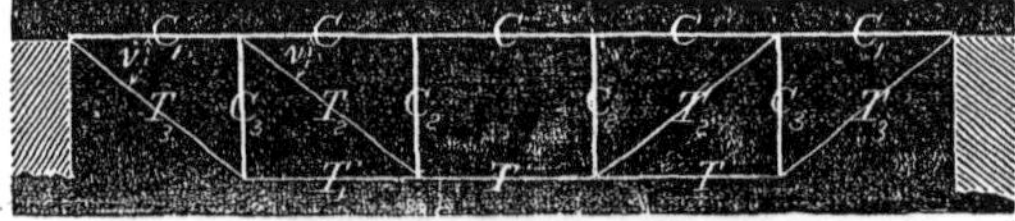

$$C = T = 3C_2 \text{ cotg. } v$$

$$C_1 = T_1 = 2C_2 \text{ cotg. } v$$

$$C_2 = \frac{W}{5}$$

$$C_3 = 2C_2$$

$$T_2 = C_2 \text{ cosec. } v$$

$$T_3 = 2T_2$$

6 Bays $= \frac{l}{6}$

Fig. 203.

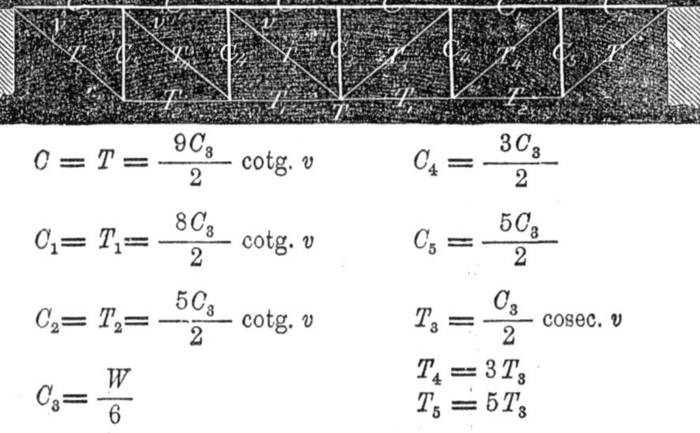

$C = T = \frac{9C_3}{2}$ cotg. v

$C_1 = T_1 = \frac{8C_3}{2}$ cotg. v

$C_2 = T_2 = \frac{5C_3}{2}$ cotg. v

$C_3 = \frac{W}{6}$

$C_4 = \frac{3C_3}{2}$

$C_5 = \frac{5C_3}{2}$

$T_3 = \frac{C_3}{2}$ cosec. v

$T_4 = 3T_3$

$T_5 = 5T_3$

TABLE OF CONSTANTS, BASED ON FOREGOING FORMULA.

Load equally distributed.

Table of constants for strains in respective member of trusses, from 2 to 6 bays, with diagonals inclined from 5° to 45°:

Reference.

W = Load in lbs., equally distributed over whole length of truss, to be multiplied by constant for strain in repective member.
v = Angle between horizontal and diagonal.
C = Compression in lbs. in respective member.
T = Tension in lbs. in respective member.

EXAMPLE.—Required, the strain in the various members of a truss of 4 bays. Length = 40 feet; load W = 80,000 lbs.; angle $v = 20°$.

Members.	Constants.	W.	Strains.
$C = T =$	$1.372 \times$	$80,000 =$	109,760 lbs.
$C_1 = T_1 =$	$1.029 \times$	$80,000 =$	82,320 "
$C_2 =$	$0.25 \times$	$80,000 =$	20,000 "
$C_3 =$	$0.375 \times$	$80,000 =$	30,000 "
$T_2 =$	$0.365 \times$	$80,000 =$	29,200 "
$T_3 =$	$1.095 \times$	$80,000 =$	87,600 "

[NOTE.—When the trusses are inverted, the strains change in kind, but not in amount.]

2 Bays $= \frac{l}{2}$

Fig. 204.

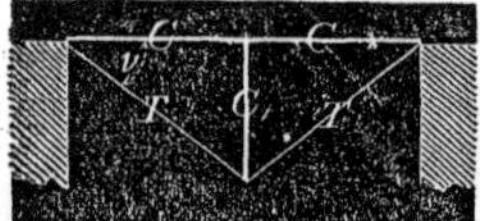

3 Bays $= \frac{l}{3}$

Fig. 205.

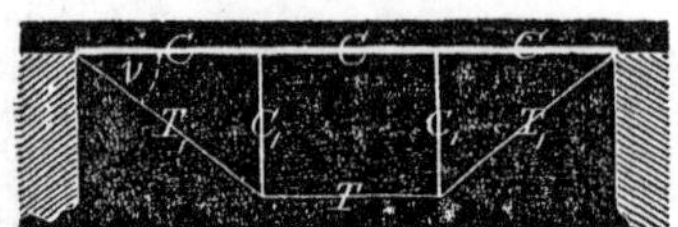

v	C	C_1	T	$C = T$	C_1	T_1
5	3.572	0.625	3.584	3.810	0.333	3.820
6	2.972	"	2 987	3.170	"	3.186
7	2.544	"	2.562	2.713	"	2.733
8	2.225	"	2.244	2.370	"	2.393
9	1.972	"	1.997	2.103	"	2.130
10	1.772	"	1 800	1.890	"	1.920
11	1.610	"	1.640	1.710	"	1.747
12	1.469	"	1.500	1.570	"	1.603
13	1.353	"	1.390	1.444	"	1.483
14	1.253	"	1.290	1.333	"	1.376
15	1.166	"	1.210	1.243	"	1.286
16	1.087	"	1.134	1.160	"	1.210
17	1.022	"	1.070	1.090	"	1.140
18	0.959	"	1.013	1.023	"	1.080
19	0.906	"	0.959	0.970	"	1.023
20	0.859	"	0.912	0.917	"	0.973
21	0.813	"	0.872	0.866	"	0.930
22	0.778	"	0.834	0.823	"	0.890
23	0.734	"	0.790	0.783	"	0.853
24	0.703	"	0.765	0.750	"	0.810
25	0.668	"	0.738	0.713	"	0.786
26	0.641	"	0.712	0.685	"	0.760
27	0.613	"	0.687	0.653	"	0.730
28	0.587	"	0.666	0.626	"	0.701
29	0.562	"	0.644	0.600	"	0.686
30	0.541	"	0.625	0.643	"	0.666
31	0.519	"	0.606	0.555	"	0.646
32	0.500	"	0.591	0.533	"	0.630
33	0.481	"	0.575	0.513	"	0.613
34	0.463	"	0.559	0.493	"	0.596
35	0.447	"	0.544	0.476	"	0.580
36	0.431	"	0.531	0.460	"	0.566
37	0.416	"	0.519	0.444	"	0.553
38	0.400	"	0.506	0.426	"	0.540
39	0.384	"	0.497	0.410	"	0.530
40	0.372	"	0.487	0.396	"	0.520
41	0.359	"	0.475	0.385	"	0.506
42	0.347	"	0.466	0.370	"	0.496
43	0.334	"	0.456	0.357	"	0.486
44	0.322	"	0.450	0.343	"	0.480
45	0.312	"	0.444	0.333	"	0.473

4 Bays $= \dfrac{l}{4}$

Fig. 206.

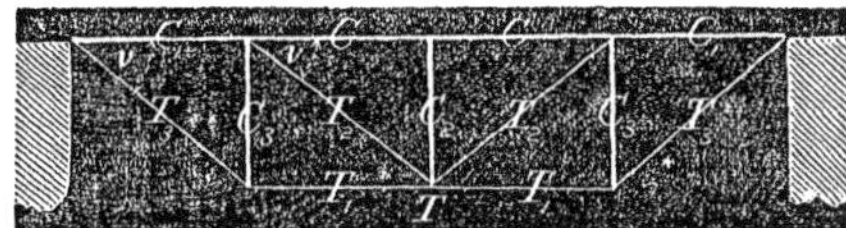

v	$C = T$	$C_1 = T_1$	C_2	C_3	T_2	T_3
5	5.720	4.290	0.250	0.375	1.434	4.032
6	4.750	3.570	"	"	1.200	3.600
7	4.068	3.051	"	"	1.025	3.075
8	3.560	2.670	"	"	0.897	2.591
9	3.164	2.373	"	"	0.799	2.397
10	2.832	2.124	"	"	0.720	2.160
11	2.568	1.926	"	"	0.655	1.965
12	2.388	1.791	"	"	0.601	1.803
13	2.164	1.623	"	"	0.556	1.668
14	2.000	1.500	"	"	0.516	1.548
15	1.864	1.398	"	"	0.482	1.446
16	1.740	1.305		"	0.454	1.362
17	1.632	1.224	"	"	0.428	1.284
18	1.532	1.149	"	"	0.405	1.215
19	1.448	1.086	"	"	0.384	1.152
20	1.372	1.029	"	"	0.365	1.095
21	1.300	0.975	"	"	0.349	1.047
22	1.236	0.927	"	"	0.334	1.002
23	1.172	0.879	"	"	0.320	0.960
24	1.124	0.843	"	"	0.306	0.918
25	1.068	0.801	"	"	0.295	0.885
26	1.024	0.768	"	"	0.285	0.855
27	0.980	0.735	"	"	0.275	0.825
28	0.940	0.705	"	"	0.266	0.798
29	0.900	0.675	"	"	0.258	0.774
30	0.864	0.648	"	"	0.250	0.750
31	0.828	0.621	"	"	0.243	0.729
32	0.800	0.600	"	"	0.236	0.708
33	0.768	0.576	"	"	0.230	0.690
34	0.740	0.555	"	"	0.224	0.672
35	0.720	0.540	"	"	0.218	0.654
36	0.688	0.516	"	"	0.212	0.636
37	0.664	0.498	"	"	0.207	0.621
38	0.640	0.480	"	"	0.203	0.609
39	0.616	0.462	"	"	0.199	0.597
40	0.600	0.450	"	"	0.195	0.585
41	0.576	0.432	"	"	0.190	0.570
42	0.560	0.420	"	"	0.186	0.558
43	0.536	0.402	"	"	0.183	0.549
44	0.520	0.390	"	"	0.180	0.540
45	0.500	0.375	"	"	0.177	0.531

$$5 \text{ Bays} = \frac{l}{5}$$

Fig. 207.

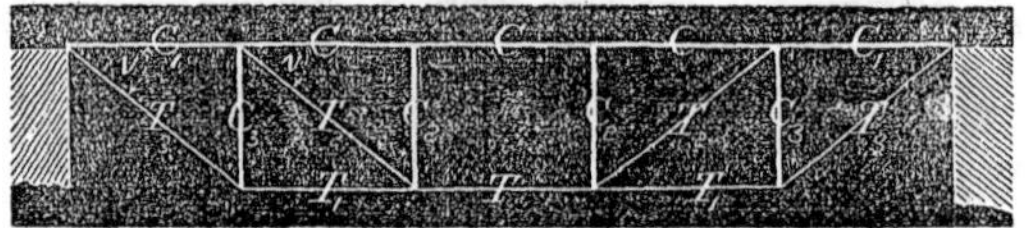

v	$C = T$	$C_1 = T_1$	C_2	C_3	T_2	T_3
5	6.858	4.572	0.200	0.400	2.294	4.588
6	5.706	3.804	"	"	1.912	3.824
7	4.884	3.256	"	"	1.640	3.280
8	4.272	2.848	"	"	1.436	2.872
9	3.786	2.524	"	"	1.278	2.556
10	3.402	2.268	"	"	1.152	2.304
11	3.084	2.056	"	"	1.048	2.096
12	2.820	1.880	"	"	0.962	1.924
13	2.598	1.732	"	"	0.890	1.780
14	2.406	1.604	"	"	0.826	1.652
15	2.238	1.492	"	"	0.772	1.544
16	2.088	1.392	"	"	0.726	1.452
17	1.962	1.308	"	"	0.684	1.368
18	1.842	1.228	"	"	0.648	1.296
19	1.740	1.160	"	"	0.614	1.228
20	1.650	1.100	"	"	0.584	1.168
21	1.560	1.040	"	"	0.558	1.116
22	1.482	0.988	"	"	0.534	1.068
23	1.410	0.940	"	"	0.512	1.024
24	1.350	0.900	"	"	0.490	0.980
25	1.284	0.856	"	"	0.472	0.944
26	1.230	0.820	"	"	0.456	0.912
27	1.176	0.784	"	"	0.440	0.880
28	1.128	0.752	"	"	0.426	0.852
29	1.080	0.720	"	"	0.412	0.824
30	1.038	0.692	"	"	0.400	0.800
31	0.996	0.664	"	"	0.388	0.776
32	0.960	0.640	"	"	0.378	0.756
33	0.924	0.616	"	"	0.368	0.736
34	0.888	0.592	"	"	0.358	0.716
35	0.858	0.572	"	"	0.348	0.696
36	0.828	0.552	"	"	0.340	0.680
37	0.798	0.532	"	"	0.332	0.664
38	0.768	0.512	"	"	0.324	0.648
39	0.738	0.492	"	"	0.318	0.636
40	0.714	0.476	"	"	0.312	0.624
41	0.690	0.460	"	"	0.304	0.608
42	0.666	0.444	"	"	0.298	0.596
43	0.642	0.428	"	"	0.292	0.584
44	0.618	0.412	"	"	0.288	0.576
45	0.600	0.400	"	"	0.284	0.568

6 Bays $= \frac{l}{6}$

Fig. 208.

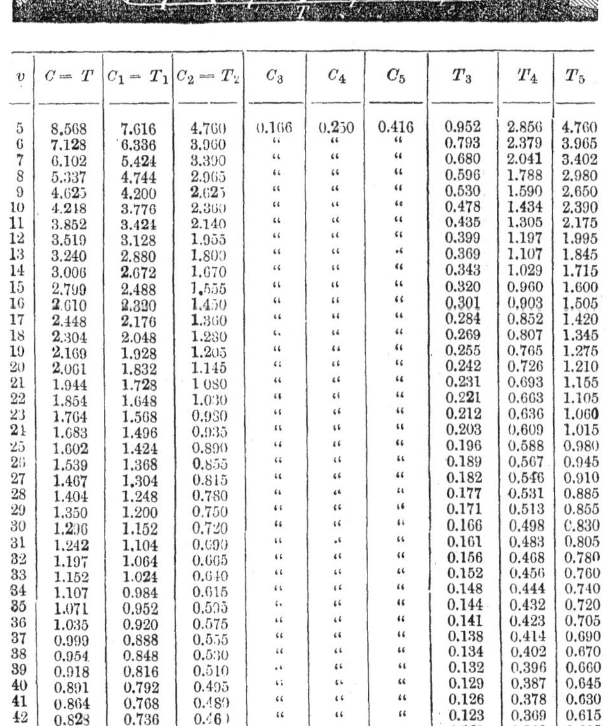

v	$C = T$	$C_1 = T_1$	$C_2 = T_2$	C_3	C_4	C_5	T_3	T_4	T_5
5	8.568	7.616	4.760	0.166	0.250	0.416	0.952	2.856	4.760
6	7.128	6.336	3.960	"	"	"	0.793	2.379	3.965
7	6.102	5.424	3.390	"	"	"	0.680	2.041	3.402
8	5.337	4.744	2.965	"	"	"	0.596	1.788	2.980
9	4.625	4.200	2.625	"	"	"	0.530	1.590	2.650
10	4.218	3.776	2.360	"	"	"	0.478	1.434	2.390
11	3.852	3.424	2.140	"	"	"	0.435	1.305	2.175
12	3.519	3.128	1.955	"	"	"	0.399	1.197	1.995
13	3.240	2.880	1.800	"	"	"	0.369	1.107	1.845
14	3.006	2.672	1.670	"	"	"	0.343	1.029	1.715
15	2.799	2.488	1.555	"	"	"	0.320	0.960	1.600
16	2.610	2.320	1.450	"	"	"	0.301	0.903	1.505
17	2.448	2.176	1.360	"	"	"	0.284	0.852	1.420
18	2.304	2.048	1.280	"	"	"	0.269	0.807	1.345
19	2.169	1.928	1.205	"	"	"	0.255	0.765	1.275
20	2.061	1.832	1.145	"	"	"	0.242	0.726	1.210
21	1.944	1.728	1.080	"	"	"	0.231	0.693	1.155
22	1.854	1.648	1.030	"	"	"	0.221	0.663	1.105
23	1.764	1.568	0.980	"	"	"	0.212	0.636	1.060
24	1.683	1.496	0.935	"	"	"	0.203	0.609	1.015
25	1.602	1.424	0.890	"	"	"	0.196	0.588	0.980
26	1.539	1.368	0.855	"	"	"	0.189	0.567	0.945
27	1.467	1.304	0.815	"	"	"	0.182	0.546	0.910
28	1.404	1.248	0.780	"	"	"	0.177	0.531	0.885
29	1.350	1.200	0.750	"	"	"	0.171	0.513	0.855
30	1.296	1.152	0.720	"	"	"	0.166	0.498	0.830
31	1.242	1.104	0.690	"	"	"	0.161	0.483	0.805
32	1.197	1.064	0.665	"	"	"	0.156	0.468	0.780
33	1.152	1.024	0.640	"	"	"	0.152	0.456	0.760
34	1.107	0.984	0.615	"	"	"	0.148	0.444	0.740
35	1.071	0.952	0.595	"	"	"	0.144	0.432	0.720
36	1.035	0.920	0.575	"	"	"	0.141	0.423	0.705
37	0.999	0.888	0.555	"	"	"	0.138	0.414	0.690
38	0.954	0.848	0.530	"	"	"	0.134	0.402	0.670
39	0.918	0.816	0.510	"	"	"	0.132	0.396	0.660
40	0.891	0.792	0.495	"	"	"	0.129	0.387	0.645
41	0.864	0.768	0.480	"	"	"	0.126	0.378	0.630
42	0.828	0.736	0.460	"	"	"	0.123	0.369	0.615
43	0.801	0.712	0.445	"	"	"	0.121	0.363	0.605
44	0.774	0.688	0.430	"	"	"	0.119	0.357	0.595
45	0.747	0.664	0.415	"	"	"	0.118	0.354	0.590

STRAINS IN TRUSSED BEAMS.

When a beam supported at the ends, is required to carry a greater load than its given capacity, and trussing is resorted to, it may become necessary to find what portion of the load is borne by the different members of the trussed beam.

Reference.

Let W = Load acting on truss at a supported point. (See figure.)
W_1 = That portion of W acting on diagonals.
W_2 = That portion of W acting on beam.
A_1 = Sectional area of diagonal.
A_2 = Sectional area of beam.
E_1 = Modulus of elasticity of material in diagonals.
E_2 = Modulus of elasticity of material in beam.
a = Length of diagonal.
b = Distance between center of beam and point of support.
c = Distance between abutment and point of support.
f = Depth of beam.
h = Depth of truss.
l = Distance between center of beam and abutment.

[NOTE.—Use the same unit of length and weight.]

No. 1.

Fig. 209

$$\frac{W_1}{W_2} = \frac{l^3}{a^3} \cdot \frac{h^2}{f^2} \cdot \frac{A_1}{A_2} \cdot \frac{E_1}{E_2}$$

$$W_1 = \frac{l^3}{a^3} \cdot \frac{h^2}{f^2} \cdot \frac{A_1}{A_2} \cdot \frac{E_1}{E_2} W_2$$

$$W_2 = \frac{a^3}{l^3} \cdot \frac{f^2}{h^2} \cdot \frac{A_2}{A_1} \cdot \frac{E_2}{E_1} W_1$$

$$A_1 = \frac{W_1}{W_2} \cdot \frac{a^3}{l^3} \cdot \frac{f^2 A_2}{h^2} \cdot \frac{E_2}{E_1}$$

$$A_2 = \frac{W_2}{W_1} \cdot \frac{l^3}{a^3} \cdot \frac{h^2 A_1}{f^2} \cdot \frac{E_1}{E_2}$$

$$W_1 = \frac{\frac{W_1}{W_2}}{\frac{W_1}{W_2} + 1} . W \qquad W_2 = \frac{}{\frac{W_1}{W_2} + 1}$$

When load is equally distributed W becomes $\frac{5}{8}$ W.

No. 2.

Fig. 210. *Fig* 211.

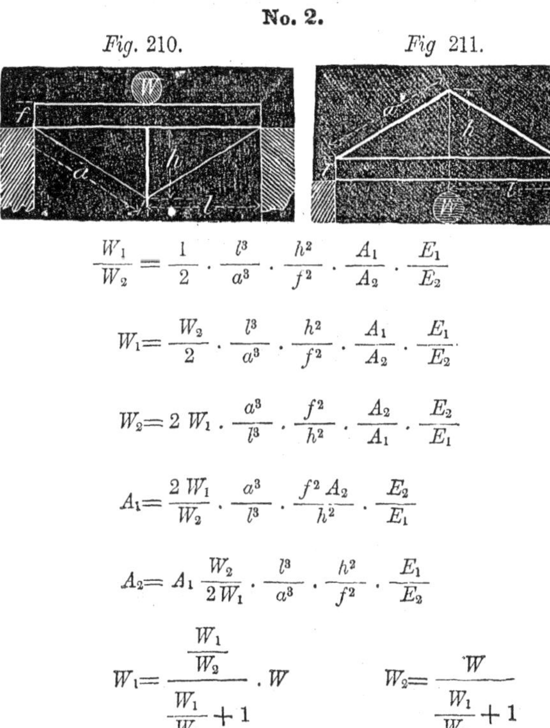

$$\frac{W_1}{W_2} = \frac{1}{2} \cdot \frac{l^3}{a^3} \cdot \frac{h^2}{f^2} \cdot \frac{A_1}{A_2} \cdot \frac{E_1}{E_2}$$

$$W_1 = \frac{W_2}{2} \cdot \frac{l^3}{a^3} \cdot \frac{h^2}{f^2} \cdot \frac{A_1}{A_2} \cdot \frac{E_1}{E_2}$$

$$W_2 = 2\, W_1 . \frac{a^3}{l^3} \cdot \frac{f^2}{h^2} \cdot \frac{A_2}{A_1} \cdot \frac{E_2}{E_1}$$

$$A_1 = \frac{2\, W_1}{W_2} \cdot \frac{a^3}{l^3} \cdot \frac{f^2 A_2}{h^2} \cdot \frac{E_2}{E_1}$$

$$A_2 = A_1 \frac{W_2}{2\, W_1} \cdot \frac{l^3}{a^3} \cdot \frac{h^2}{f^2} \cdot \frac{E_1}{E_2}$$

$$W_1 = \frac{\frac{W_1}{W_2}}{\frac{W_1}{W_2} + 1} . W \qquad W_2 = \frac{W}{\frac{W_1}{W_2} + 1}$$

When load is equally distributed W becomes $\frac{5}{8}$ W.

No. 3.

Fig. 212.

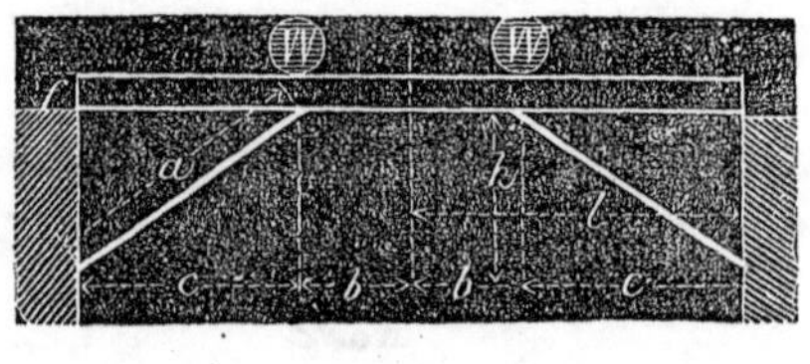

$$\frac{W_1}{W_2}=\frac{h^2}{f^2}\cdot\frac{(l^2-b^2)\,c}{a\,(a^2+bc)}\cdot\frac{A_1}{A_2}\cdot\frac{E_1}{E_2}$$

$$W_1=\frac{h^2}{f^2}\cdot\frac{(l^2-b^2)\,c}{a\,(a^2\times bc)}\cdot\frac{A_1}{A_2}\cdot\frac{E_1}{E_2}\,.\,W_2$$

$$W_2=\frac{f^2}{h^2}\cdot\frac{a\,(a^2+bc)}{(l^2-b^2)\,c}\cdot\frac{A_2}{A_1}\cdot\frac{E_2}{E_1}\,.\,W_1$$

$$A_1=\frac{W_1}{W_2}\cdot\frac{f^2}{h^2}\cdot\frac{a\,(a^2+bc)}{(l^2-b^2)\,c}\cdot\frac{A_2\,.\,E_2}{E_1}$$

$$A_2=\frac{W_2}{W_1}\cdot\frac{h^2}{f^2}\cdot\frac{(l^2-b^2)\,c}{a\,(a^2+bc)}\cdot\frac{A_1\,.\,E_1}{E_2}$$

$$W_1=\frac{\frac{W_1}{W_2}}{\frac{W_1}{W_2}+1}\,.\,W \qquad W_2=\frac{W}{\frac{W_1}{W_2}+1}$$

When load is equally distributed W becomes $\frac{5}{8}\,W$.

No. 4.

Figs. 213 and 214.

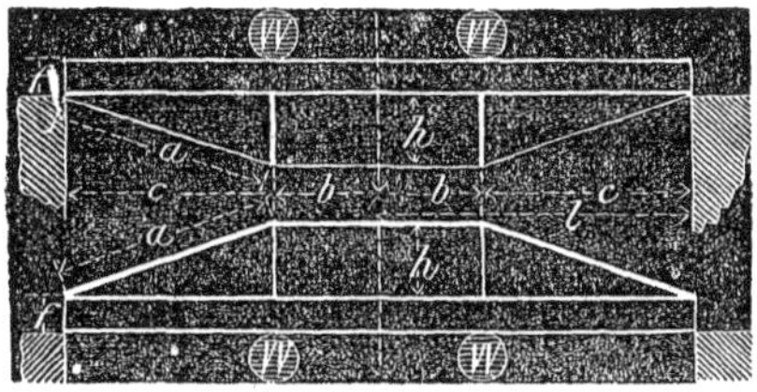

$$\frac{W_1}{W_2} = \frac{h^2}{2f^2} \cdot \frac{(l^2 - b^2)\,c}{a\,(a^2 + bc)} \cdot \frac{A_1}{A_2} \cdot \frac{E_1}{E_2}$$

$$W_1 = \frac{h^2}{2f^2} \cdot \frac{(l^2 - b^2)\,c}{a\,(a^2 + bc)} \cdot \frac{A_1}{A_2} \cdot \frac{E_1}{E_2} \cdot W$$

$$W_2 = 2\,W_1 \frac{f^2}{h^2} \cdot \frac{a\,(a^2 + bc)}{(l^2 - b^2)\,c} \cdot \frac{A_2}{A_1} \cdot \frac{E_2}{E_1}$$

$$A_1 = \frac{2\,W_1}{W_2} \cdot \frac{f^2}{h^2} \cdot \frac{a\,(a^2 + bc)}{(l^2 - b^2)\,c} \cdot \frac{A_2 \,.\, E_2}{E_1}$$

$$A_2 = \frac{W_2}{2\,W_1} \cdot \frac{h^2}{f^2} \cdot \frac{(l^2 - b^2)\,c}{a\,(a^2 + bc)} \cdot \frac{A_1 \,.\, E_1}{E_2}$$

$$W_1 = \frac{\frac{W_1}{W_2}}{\frac{W_1}{W_2} + 1} \cdot W \qquad W_2 = \frac{W}{\frac{W_1}{W_2} + 1}$$

When load is equally distributed W becomes $\frac{3}{8}\,W$.

STRAINS IN TRUSSES, WITH PARALLEL BOOMS.

(*Caused by Static and Moving Loads.*)

The strain in the upper boom is always compressive.

The strain in the lower boom is always tensile.

All braces inclined *down* from the nearest abutment are in tension.

All braces inclined *up* from the nearest abutment are in compression.

The strains in the verticals and diagonals *increase* from the center of truss to abutment.

The strains in the booms *decrease* from the center of truss to abutment.

A moving load, advancing over a truss, &c., causes the maximum moment of rupture (which under an equally distributed load is at the center of truss) to shift to one side of the center, thereby changing the nature and amount of strain in web only. This requires either the enlargement of those members constituting the web or the addition of so-called counters, (braces, struts, or ties.)

To find the point from center of truss to where the addition of counters must commence, the following formula is used:

Let d = Distance from center of truss to point where maximum moment of rupture occurs, and where counter bracing must commence.

d' = Distance from nearest abutment to ditto.

$$\text{Then will } d = l\left[\frac{1}{2} + \frac{w}{w_{\prime}} - \sqrt{\frac{w}{w_{\prime}}\left(1 + \frac{w}{w_{\prime}}\right)}\,\right]$$

$$\text{And } d_{\prime} = \frac{l}{2} - d = \frac{lw}{w_{\prime}}\left[\left(\sqrt{1 + \frac{w_{\prime}}{w}}\right) - 1\right]$$

These results will be found to agree with formulas for "Counter Strains" when V_m becomes negative.

Reference.

N = Total number of bays in a truss.

H_n = Horizontal strains in booms.

V_n = Strains in verticals.

Y_n = Strains in diagonals.

V_m = Vertical strains acting on counters Y_m.

Y_m = Strains in counters, opposite in kind to Y_n.

W = Weight of static load, equally distributed over whole length of truss.

$W_{/}$ = Weight of moving load, equally distributed over whole length of truss.

h = Height or depth of truss between the center of gravity of booms.

l = Span or length of truss from abutment to abutment.

n = Number of member, counting from abutment A.

m = Number of member, between center and abutment B.

r = Half the length of a panel or bay.

s = Length of a panel or bay.

w = Weight of static load per unit of length l.

$w_{/}$ = Weight of moving load per unit of length l.

v = Angle between horizontal and diagonal.

For other designations, see diagrams and examples.

The angle v for Howe Truss is generally 45°.
The angle v for Whipple Truss is generally 45°.
The angle v for Lattice Truss is generally 45°.
The angle v for Warren Truss is generally 60°.

The proportion of height h to span l is from $\frac{1}{7}$ to $\frac{1}{15}$, generally $\frac{1}{10}$.

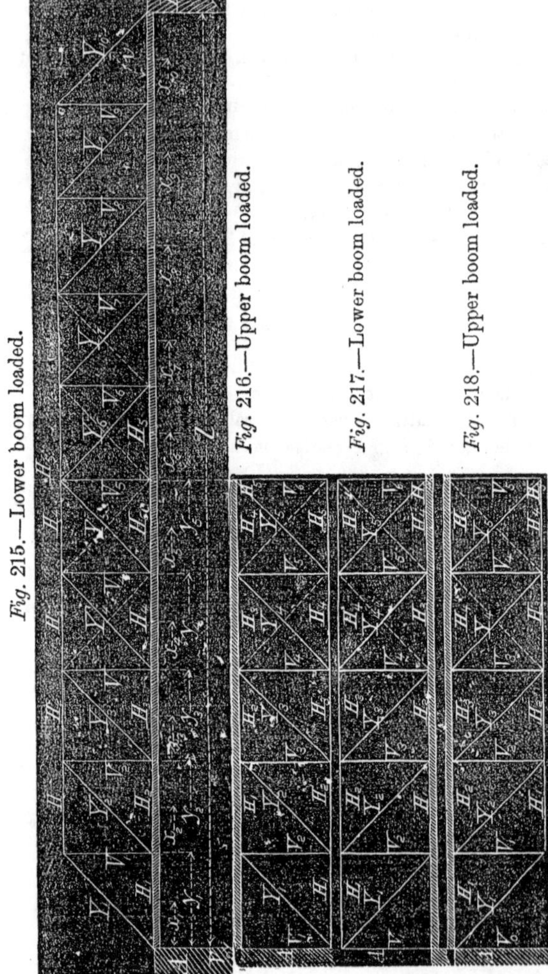

Fig. 215.—Lower boom loaded.

Fig. 216.—Upper boom loaded.

Fig. 217.—Lower boom loaded.

Fig. 218.—Upper boom loaded.

HOWE TRUSS. (*Figs.* 215, 216, 217, and 218.)

Additional Reference.

x_n= Distance from abutment A to center of bay.
y_n= Distance from abutment A to apex of bay.

Static or Permanent Load, equally distributed over whole length of Truss.

Strains in Booms.

$$H_n = \frac{W}{2h} \cdot y_n - \frac{W}{2hl} \cdot y_n^2$$

Strains in Verticals.

$$V_n = \frac{W}{2} - \frac{W}{l} x_n$$

Strains in Diagonals.

$$Y_n = V_n \text{ cosec. } v.$$

Moving and Static Load, each equally distributed per unit of length.

Strains in Booms.

$$H_n = \frac{W + W_1}{2h} \cdot y_n - \frac{W + W_1}{2hl} \cdot y_n^2$$

Strains in Verticals.

$$V_n = \frac{W}{2} - \frac{W}{l} x_n + \frac{W_1}{2l^2} (l - x_n)^2$$

Strains in Diagonals.

$$Y_n = V_n \text{ cosec. } v.$$

Strains in Counters.

$$V_m = \frac{W}{2} - \frac{W}{l} x_m + \frac{W_1}{2l^2} (l - x_m)^2 \qquad Y_m = V_m \text{ cosec. } v.$$

EXAMPLE. (*Figs.* 215, 216, 217, and 218.)

Moving Load, (as railway train passing over bridge.)

We will assume $W = 50,000$ lbs.
$W_1 = 100,000$ lbs.
$l = 100$ feet.
$h = 10$ feet.
$v = 45°$, (cosec. $= 1.414$.)

Horizontal Strains in Booms, (compression in upper, tension in lower.)

$$H_n = \frac{W + W_1}{2h} . y_n - \frac{W + W_1}{2hl} . y_n^2 = \frac{50000 + 100000}{20} .$$

$$y_n - \frac{50000 + 100000}{2000} . y_n^2 = 7500 . y_n - 75 . y_n^2$$

$H_1 = 7500.10 - 75.100 = 67,500$ lbs.
$H_2 = 7500.20 - 75.400 = 120,000$ lbs.
$H_3 = 7500.30 - 75.900 = 157,500$ lbs.
$H_4 = 7500.40 - 75.1600 = 180,000$ lbs.
$H_5 = 7500.50 - 75.2500 = 187,500$ lbs.

Strains in Verticals.

$$V_n = \frac{W}{2} - \frac{W}{l} . x_n + \frac{W_1}{2l^2} (l - x_n)^2 = \frac{50000}{2} - \frac{50000}{100} .$$

$$x_n + \frac{100000}{20000} . (l - x_n)^2 = 25000 - 500 . x_n + 5 (l - x_n)^2$$

	Strains in Figs. 215	216	217	218
$V_1 = 25000 - 500.5 + 5.95^2 = 67625$	Ten.	Ten.	Com.	Com.
$V_2 = 25000 - 500.15 + 5.85^2 = 53625$	"	"	"	"
$V_3 = 25000 - 500.25 + 5.75^2 = 40625$	"	"	"	"
$V_4 = 25000 - 500.35 + 5.65^2 = 28625$	"	"	"	"
$V_5 = 25000 - 500.45 + 5.55^2 = 17625$	"	"	"	"

Counter Strains (V_m) *for Strains in Counters.*

$V_6 = 25000 - 500.55 + 5.45^2 = 7625.$
$V_7 = 25000 - 500.65 + 5.35^2 = 5625.$

Strains in Diagonals.

$Y_n = V_n$ cosec v.

	Strains in Figs. 215	216	217	218
$Y_1 = 67625 . 1.414 = 95,620$ lbs.	Com.	Com.	Ten.	Ten.
$Y_2 = 53625 . 1.414 = 75,826$ lbs.	"	"	"	"
$Y_3 = 40625 . 1.414 = 57,441$ lbs.	"	"	"	"
$Y_4 = 28625 . 1.414 = 40,476$ lbs.	"	"	"	"
$Y_5 = 17625 . 1.414 = 24,922$ lbs.	"	"	"	"

Strains in Counters, (dotted lines, Fig. 215, *for example.)*

$Y_m = V_m$ cosec. v.

	Strains in Figs. 215	216	217	218
$Y_6 = 7625 . 1.414 = 10,782$ lbs.	Com.	Com.	Ten.	Ten.
$Y_7 = 5625 . 1.414 = 7,954$ lbs.	"	"	"	"

Fig. 219.

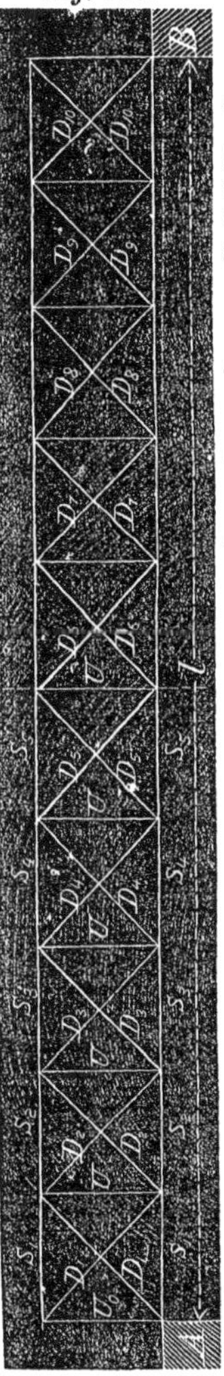

LATTICE TRUSS WITH VERTICAL NUMBERS.

Fig. 219. *Load on either Boom.*

To compute the strains in this truss, the easiest method is to find the values of H_n, V_n, V_m, Y_n, and Y_m for a Howe Truss, (*Figs.* 215, 216, 217, and 218,) loaded in the same manner, (upper or lower boom.) These values in the following formulas for the above truss will give the required strains:

Strains in Booms. (*S.*)

$$S_1 = \frac{H_1}{2} \qquad S_4 = \frac{H_3 + H_4}{2}$$

$$S_2 = \frac{H_1 + H_2}{2} \qquad S_5 = \frac{H_4 + H_5}{2}$$

$$S_3 = \frac{H_2 + H_3}{2} \quad \text{Generally } S_n = \frac{H_{n-1} + H_n}{2}$$

Strains in Verticals. (*U.*)

Upper boom loaded—compression.
Lower boom loaded—tension.

$$U = \frac{W + W_1}{2N} \quad \text{constant.}$$

Strains in End Post (U_o.)

Upper boom loaded.
$U_o = U + S_1 =$ compression.

Lower boom loaded.
$U_o = S_1 =$ compression.

Strains in Diagonals. (*D.*)

$$D_1 = \frac{Y_1}{2} \qquad D_4 = \frac{Y_4}{2}$$

$$D_2 = \frac{Y_2}{2} \qquad D_5 = \frac{Y_5}{2}$$

$$D_3 = \frac{Y_3}{2} \quad \text{Generally } D_n = \frac{Y_n}{2}$$

Strains in Counters.

$$\text{Generally } D_m = \frac{Y_m}{2}$$

Fig. 220.

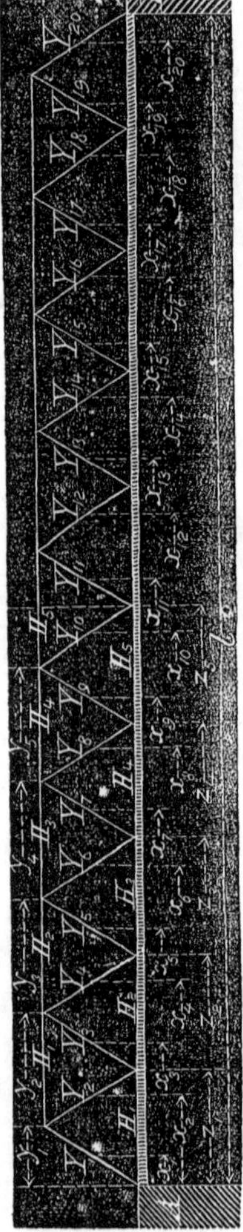

WARREN TRUSS.

Fig. 220. *Lower Boom Loaded.*

Additional Reference.

x_n = Distance from abutment A to center of diagonal.

y_n = Distance from abutment A to apex of bay of upper boom.

z_n = Distance from abutment A to apex of bay of lower boom.

Static or Permanent Load, equally distributed over whole length of Truss.

Strains in Booms.

Upper.

$$H_n = .\frac{W}{2h} z_n - \frac{W}{2hl} . z_n^2$$

Lower.

$$H_n = \frac{W}{2h} . y_n - \frac{W}{2hl} . y_n^2$$

Strains in Verticals.

$$V_n = \frac{W}{2} - \frac{W}{l} x_n$$ (V_n acts at the end of x_n.)

Strains in Diagonals.

$$Y_n = V_n \text{ cosec. } v.$$

Moving and Static Load, each equally distributed per unit of length.

Strains in Booms.

Upper.

$$H_n = \frac{W + W_1}{2h} . z_n - \frac{W + W_1}{2hl} . z_n^2$$

Lower.

$$H_n = \frac{W + W_1}{2h} . y_n - \frac{W + W_1}{2hl} y_n^2$$

Strains in Verticals.

$$V_n = \frac{W}{2} - \frac{W}{l} x_n + \frac{W_1}{2l^2} (l - x_n)^2$$

Strains in Diagonals.

$$Y_n = V_n \text{ cosec } v.$$

Strains in Counters.

$$V_m = \frac{W}{2} - \frac{W}{l} x_m + \frac{W_1}{2l^2}(l - x_m)^2 \qquad Y_m = V_m \text{ cosec. } v.$$

EXAMPLE. (*Fig.* 220.)

Moving Load (as railway train passing over bridge) on lower Boom.

We will assume $W = 50{,}000$ lbs.
$W_1 = 100{,}000$ lbs.
$l = 100$ feet.
$h = 10$ feet.
$v = 63° \, 20'$, (cosec. $= 1.12$.)

Horizontal Strains in Upper Boom. (*Compression.*)

$$H_n = \frac{W + W_1}{2h} . z_n - \frac{W + W_1}{2hl} . z_n^2 = \frac{50000 + 100000}{2.10}$$

$$z_n - \frac{50000 + 100000}{2.10 \,.\, 100} . z_n^2 = \frac{150000}{20} . z_n -$$

$$\frac{150000}{2000} z_n^2 = 7500 . z_n - 75 . z_n^2$$

$H_1 = 7500.10 - 75.100 = 67{,}500$ lbs.
$H_2 = 7500.20 - 75.400 = 120{,}000$ lbs.
$H_3 = 7500.30 - 75.900 = 157{,}500$ lbs.
$H_4 = 7500.40 - 75.1600 = 180{,}000$ lbs.
$H_5 = 7500.50 - 75.2500 = 187{,}500$ lbs.

Horizontal Strains in Lower Boom. (*Tension.*)

$$H_n = \frac{W + W_1}{2h} . y_n - \frac{W + W_1}{2hl} . y_n^2 = \frac{50000 + 100000}{2.10} .$$

$$y_n - \frac{50000 + 100000}{2.10 \,.\, 100} . y_n^2 = \frac{150000}{20} . y_n - \frac{150000}{2000} . y_n^2$$

$H_1 = 7500.5 - 75.25 = 37500 - 1875 = 35{,}625$ lbs.
$H_2 = 7500.15 - 75.225 = 112500 - 16875 = 95{,}625$ lbs.
$H_3 = 7500.25 - 75.625 = 187500 - 46875 = 140{,}625$ lbs.
$H_4 = 7500.35 - 75.1225 = 262500 - 91875 = 170{,}625$ lbs.
$H_5 = 7500.45 - 75.2025 = 337500 - 151875 = 185{,}625$ lbs.

Strains in Verticals.

$$Y_n = V_n \text{ cosec. } v.$$

$$V_n = \frac{W}{2} - \frac{W}{l}.x_n + \frac{W_1}{2l}.(l - x_n) = \frac{50000}{2} - \frac{50000}{100}.$$

$$x_n + \frac{100000}{2.100^2}.(100 - x_n)^2 = 25000 - 500x_n + 5.(100 - x_n)^2$$

$V_1 = 25000 - 500 . 2.5 + 5 . 9506.25 = 71281.25.$
$V_2 = 25000 - 500 . 7.5 + 5 . 8556.25 = 64031.25.$
$V_3 = 25000 - 500 . 12.5 + 5 . 7656.25 = 57031.25.$
$V_4 = 25000 - 500 . 17.5 + 5 . 6806.25 = 50281.25.$
$V_5 = 25000 - 500 . 22.5 + 5 . 6006.25 = 43781.25.$
$V_6 = 25000 - 500 . 27.5 + 5 . 5256.25 = 37531.25.$
$V_7 = 25000 - 500 . 32.5 + 5 . 4556.25 = 31531.25.$
$V_8 = 25000 - 500 . 37.5 + 5 . 3906.25 = 25781.25.$
$V_9 = 25000 - 500 . 42.5 + 5 . 3306.25 = 20281.25.$
$V_{10} = 25000 - 500 . 47.5 + 5 . 2756.25 = 14031.25.$

Counter Strains. $(V_{m'})$

$V_{11} = 25000 - 500 . 52.5 + 5 . 2256.25 = 10031.25.$
$V_{12} = 25000 - 500 . 57.5 + 5 . 1806.25 = 5281.25.$
$V_{13} = 25000 - 500 . 62.5 + 5 . 1406.25 = 781.25.$
$V_{14} =$ Null.

Strains in Diagonals.

$$Y_n = V_n \text{ cosec. } v.$$

$Y_1 = 71281.25 . 1.12 = 79,835$ lbs. Compression in Y_1 and Y_{20}.
$Y_2 = 64031.25 . 1.12 = 71,715$ lbs. Tension in Y_2 and Y_{19}.
$Y_3 = 57031.25 . 1.12 = 63,875$ lbs. Compression in Y_3 and Y_{18}.
$Y_4 = 50281.25 . 1.12 = 56,315$ lbs. Tension in Y_4 and Y_{17}.
$Y_5 = 43781.25 . 1.12 = 49,035$ lbs. Compression in Y_5 and Y_{16}.
$Y_6 = 37531.25 . 1.12 = 42,035$ lbs. Tension in Y_6 and Y_{15}.
$Y_7 = 31531.25 . 1.12 = 35,315$ lbs. Compression in Y_7 and Y_{14}.
$Y_8 = 25781.25 . 1.12 = 28,875$ lbs. Tension in Y_8 and Y_{13}.
$Y_9 = 20281.25 . 1.12 = 22,715$ lbs. Compression in Y_9 and Y_{12}.
$Y_{10} = 14031.25 . 1.12 = 15,715$ lbs. Tension in Y_{10} and Y_{11}.

Counter Strains.

$$Y_m = V_m \text{ cosec. } v.$$

$Y_{11} = 10031.25 . 1.12 = 11,235$ lbs. Compression in Y_{10} and Y_{11}.
$Y_{12} = 5281.25 . 1.12 = 5,915$ lbs. Tension in Y_9 and Y_{12}.
$Y_{13} = 781.25 . 1.12 = 875$ lbs. Compression in Y_8 and Y_{13}.

Fig. 221.

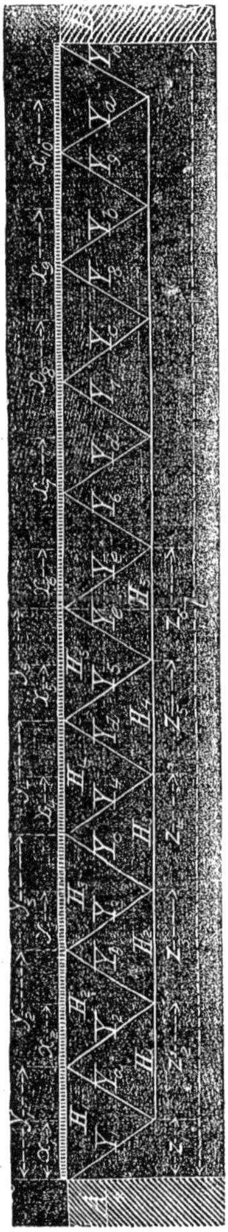

WARREN TRUSS.

Fig. 221. *Upper Boom Loaded.*

Additional Reference.

x_n = Distance from abutment A to center of bay of upper boom.

y_n = Distance from abutment A to apex of bay of upper boom.

z_n = Distance from abutment A to apex of bay of lower boom.

Static or Permanent Load, equally distributed over whole length of Truss.

Strains in Booms.

Upper.

$$H_n = \frac{W}{2h} \cdot z_n - \left(\frac{W}{2hl} \cdot z_n{}^2 + \frac{W r^2}{2hl} \right)$$

Lower.

$$H_n = \frac{W}{2h} \cdot y_n - \frac{W}{2hl} \cdot y_n^2$$

Strains in Verticals.

$$V_n = \frac{W}{2} - \frac{W}{l} \cdot x_n$$

Strains in Diagonals.

$$Y_n = V_n \text{ cosec. } v.$$

Moving and Static Load, each equally distributed per unit of length.

Strains in Booms.

Upper.

$$H_n = \frac{W + W_1}{2h} \cdot z_n - \left(\frac{W + W_1}{2hl} \cdot z_n^2 + \frac{(W + W_1)\, r^2}{2hl} \right)$$

Lower.

$$H_n = \frac{W + W_1}{2h} \cdot y_n - \frac{W + W_1}{2hl} \cdot y_n^2$$

Strains in Verticals.

$$V_n = \frac{W}{2} - \frac{W}{l} x_n + \frac{W_1}{2l^2} (l - x_n)^2$$

Strains in Diagonals.

$$Y_n = V_n \text{ cosec. } v.$$

Strains in Counters.

$$V_m = \frac{W}{2} - \frac{W}{l} x_m + \frac{W_1}{2l^2} (l - x_m)^2 \qquad Y_m = V_m \text{ cosec. } v.$$

EXAMPLE. (*Fig.* 221.)

Moving Load (as railway train passing over bridge) on Upper Boom.

We will assume $W = 50{,}000$ lbs.
$W_1 = 100{,}000$ lbs.
$l = 100$ feet.
$h = 10$ feet.
$v = 63° \ 20'$, $r = 5$ feet.

Horizontal Strains in Upper Boom. (Compression.)

$$H_n = \frac{W + W_1}{2h} . z_n - \left[\frac{W + W_1}{2hl} . z_n^2 + \frac{(W + W_1) r^2}{2hl} \right] =$$

$$\frac{150000}{20} . z_n - \left[\frac{150000}{2000} . z_n^2 + \frac{150000 . 5^2}{2000} \right] =$$

$$7500 . z_n - [75 . z_n^2 + 1875]$$

$$H_1 = 7500.5 \ - [75.25 \ + 1875] = \ 33{,}750 \text{ lbs.}$$
$$H_2 = 7500.15 - [75.225 \ + 1875] = \ 93{,}750 \text{ lbs.}$$
$$H_3 = 7500.25 - [75.625 \ + 1875] = 138{,}750 \text{ lbs.}$$
$$H_4 = 7500.35 - [75.1225 + 1875] = 168{,}750 \text{ lbs.}$$
$$H_5 = 7500.45 - [75.2025 + 1875] = 183{,}750 \text{ lbs.}$$

Horizontal Strains in Lower Boom (Tension.)

$$H_n = \frac{W + W_1}{2h} . y_n - \frac{W + W_1}{2hl} . y_n^2 = 7500 . y_n - 75 . y_n^2$$

$$H_1 = 7500.10 - 75.100 \ = \ 67{,}500 \text{ lbs.}$$
$$H_2 = 7500.20 - 75.400 \ = 120{,}000 \text{ lbs.}$$
$$H_3 = 7500.30 - 75.900 \ = 157{,}500 \text{ lbs.}$$
$$H_4 = 7500.40 - 75.1600 = 180{,}000 \text{ lbs.}$$
$$H_5 = 7500.50 - 75.2500 = 187{,}500 \text{ lbs.}$$

Strains in Verticals.

$$V_n = \frac{W}{2} - \frac{W}{l}.x_n + \frac{W_1}{2l^2}(l - x_n)^2 = 25000 - 500.x_n + 5.(l - x_n)^2$$

$V_1 = 25000 - 500.5 + 5.95^2 = 67{,}625$ lbs.
$V_2 = 25000 - 500.15 + 5.85^2 = 53{,}625$ lbs.
$V_3 = 25000 - 500.25 + 5.75^2 = 40{,}625$ lbs.
$V_4 = 25000 - 500.35 + 5.65^2 = 28{,}625$ lbs.
$V_5 = 25000 - 500.45 + 5.55^2 = 17{,}625$ lbs.

Counter Strains.

$V_6 = 25000 - 500.55 + 5.45^2 = 7{,}625$ lbs.

Strains in Diagonals.

$$Y_n = V_n \text{ cosec.}$$

$Y_1 = 67625 \,.\, 1.12 = 75{,}740$ lbs. Tension in Y_1 and Y_{10}; compression in Y_a and Y_a.
$Y_2 = 53625 \,.\, 1.12 = 60{,}060$ lbs. Tension in Y_2 and Y_9; compression in Y_b and Y_b.
$Y_3 = 40625 \,.\, 1.12 = 45{,}500$ lbs. Tension in Y_3 and Y_8; compression in Y_c and Y_c.
$Y_4 = 28625 \,.\, 1.12 = 32{,}060$ lbs. Tension in Y_4 and Y_7; compression in Y_d and Y_d.
$Y_5 = 17625 \,.\, 1.12 = 19{,}740$ lbs. Tension in Y_5 and Y_6; compression in Y_e and Y_e.

Counter Strains.

$$Y_m = V_m \text{ cosec. } v.$$

$Y_6 = 7625 \,.\, 1.12 = 8{,}540$ lbs. Compression in Y_5 and Y_6; tension in Y_e and Y_e.

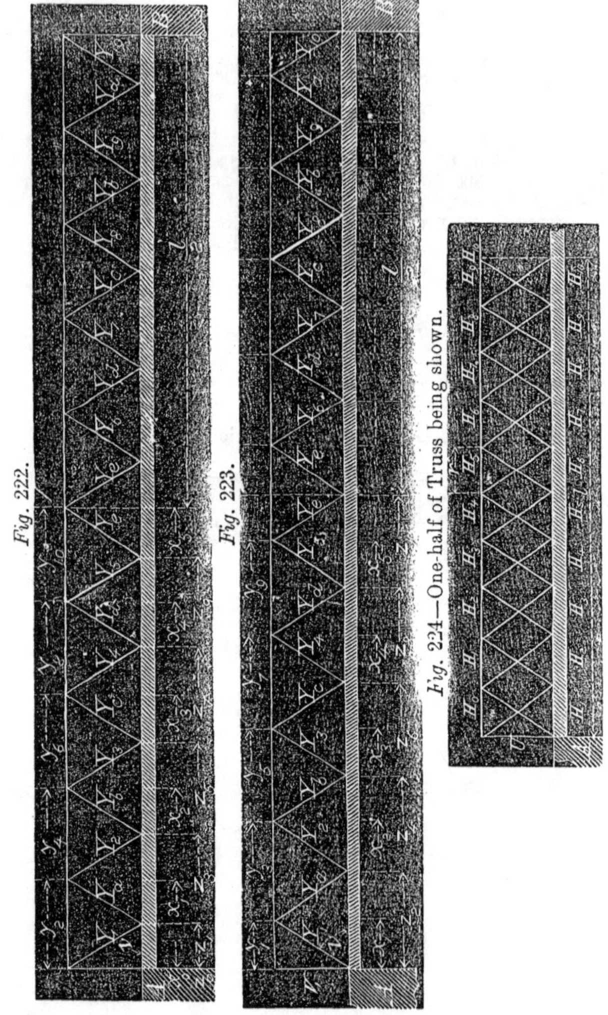

Fig. 222.

Fig. 223.

Fig. 224—One-half of Truss being shown.

LATTICE TRUSS. (*Figs.* 222, 223, and 224.)

Lower Boom Loaded.

Additional Reference.

$r=$ Half the length of a bay of *simple* truss. (*Figs.* 222 and 223.)
$x_n=$ Distance from abutment A to center of bay of *lower* boom.
$y_n=$ Distance from abutment A to apex of bay of *upper* boom.
$z_n=$ Distance from abutment A to apex of bay of *lower* boom.

The formulas are for the strains in the simple trusses, (*Figs.* 222 and 223.) *Fig.* 224 shows the simple trusses combined, constituting the Lattice Truss.

When the upper boom is loaded, treat the strains as acting upward and the truss inverted: the strains will be of the same amount in each member, but different in kind.

Static or Permanent Load, equally distributed over whole length of Truss.

Strains in Booms.

Upper.

$$H_n=\frac{W}{2h}\cdot\left(z_n+\frac{r}{2}\right)-\frac{W}{2hl}\cdot\left(z_n+\frac{r}{2}\right)^2+\frac{Wr^2}{8hl}$$

Lower.

$$H_n=\frac{W}{2h}\cdot\left(y_n-\frac{r}{2}\right)-\frac{W}{2hl}\cdot\left(y_n-\frac{r}{2}\right)^2-\frac{3\,Wr^2}{8hl}$$

Strains in Verticals.

$$V_n=\frac{W}{4}-\frac{W}{2l}\cdot x_n$$

Strains in Diagonals.

$$Y_n=V_n\ \text{cosec.}\ v.$$

Moving and Static Load, each equally distributed per unit of length.

Strains in Booms.

Upper.

$$H_n=\frac{W+W_1}{2h}\cdot\left(z_n+\frac{r}{2}\right)-\frac{W+W_1}{2hl}\cdot\left(z_n+\frac{r}{2}\right)^2+\frac{(W+W_1)r^2}{8hl}$$

Lower.

$$H_n=\frac{W+W_1}{2h}\cdot\left(y_n-\frac{r}{2}\right)-\frac{W+W_1}{2hl}\cdot\left(y_n-\frac{r}{2}\right)^2-\frac{3(W+_1)rW^2}{8hl}$$

Strains in Verticals.

$$V_n = \frac{W}{4} - \frac{W}{2l} \cdot x_n + \frac{W_1}{4l^2} \cdot (l - x_n)^2$$

Strains in Diagonals.

$$Y_n = V_n \text{ cosec } v.$$

Strains in Counters.

$$V_m = \frac{W}{4} - \frac{W}{2l} \cdot x_m + \frac{W_1}{4l^2} \cdot (l - x_m)^2 \qquad Y_m = V_m \text{ cosec. } v.$$

[NOTE.—The strains in $Y_{a, b, c}$, ... are equal in amount, but different in kind to the strains in $Y_{1, 2, 3}$, ...

EXAMPLE. (*Figs.* 222, 223, and 224.)

Moving Load (as railway train passing over bridge) on Lower Boom.

We will assume $W = 50{,}000$ lbs.
$W_1 = 100{,}000$ lbs.
$l = 100$ feet.
$h = 10$ feet.
$v = 63° 20'$, (cosec. $= 1.12$,) $r = 5$ feet..

Horizontal Strains in Upper Boom. (Compression. Fig. 224.)

$$H_n = \frac{W + W_1}{2h} \left(z_n + \frac{r}{2}\right) - \frac{W + W_1}{2hl} \left(z_n + \frac{r}{2}\right)^2 +$$

$$\frac{(W + W_1)\, r^2}{8hl} = 7500\,(z_n + 2.5) - 75\,(z_n + 2.5)^2 + 468.75$$

$H_0 = 7500 \,.\, (\;0 + 2.5) - 75 \,.\, (\;0 + 2.5)^2 + 468.75 = \;\;18{,}750$ lbs.
$H_1 = 7500 \,.\, (\;5 + 2.5) - 75 \,.\, (\;5 + 2.5)^2 + 468.75 = \;\;52{,}500$ lbs.
$H_2 = 7500 \,.\, (10 + 2.6) - 75 \,.\, (10 + 2.5)^2 + 468.75 = \;\;82{,}500$ lbs.
$H_3 = 7500 \,.\, (15 + 2.5) - 75 \;\; (15 + 2.5)^2 + 468.75 = 108{,}750$ lbs.
$H_4 = 7500 \,.\, (20 + 2.5) - 75 \,.\, (20 + 2.5)^2 + 468.75 = 131{,}250$ lbs.
$H_5 = 7500 \,.\, (25 + 2.5) - 75 \,.\, (25 + 2.5)^2 + 468.75 = 150{,}000$ lbs.
$H_6 = 7500 \,.\, (30 + 2.5) - 75 \,.\, (30 + 2.5)^2 + 468.75 = 165{,}000$ lbs.
$H_7 = 7500 \,.\, (35 + 2.5) - 75 \,.\, (35 + 2.5)^2 + 468.75 = 176{,}250$ lbs.
$H_8 = 7500 \,.\, (40 + 2.5) - 75 \,.\, (40 + 2.5)^2 + 468.75 = 183{,}750$ lbs.
$H_9 = 7500 \,.\, (45 + 2.5) - 75 \,.\, (45 + 2.5)^2 + 458.75 = 187{,}500$ lbs.

Horizontal Strains in Lower Boom. (*Tension. Fig.* 224.)

$$H_n = \frac{W + W_1}{2h} \cdot \left(y_n - \frac{r}{2}\right) - \frac{W + W_1}{2hl} \cdot \left(y_n - \frac{r}{2}\right)^2 -$$

$$\frac{3(W + W_1)r^2}{8hl} = 7500 . (y_n - 2.5) - 75 . (y_n - 2.5)^2 - 1406.25$$

$H_1 = 7500 . (5 - 2.5) - 75 . (5 - 2.5)^2 - 1406.25 = 16,875$ lbs.
$H_2 = 7500 . (10 - 2.5) - 75 . (10 - 2.5)^2 - 1406.25 = 50,625$ lbs.
$H_3 = 7500 . (15 - 2.5) - 75 . (15 - 2.5)^2 - 1406.25 = 80,625$ lbs.
$H_4 = 7500 . (20 - 2.5) - 75 . (20 - 2.5)^2 - 1406.25 = 106,875$ lbs.
$H_5 = 7500 . (25 - 2.5) - 75 . (25 - 2.5)^2 - 1406.25 = 129,375$ lbs.
$H_6 = 7500 . (30 - 2.5) - 75 . (30 - 2.5)^2 - 1406.25 = 148.125$ lbs.
$H_7 = 7500 . (35 - 2.5) - 75 . (35 - 2.5)^2 - 1406.25 = 163,125$ lbs.
$H_8 = 7500 . (40 - 2.5) - 75 . (40 - 2.5)^2 - 1406.25 = 174,375$ lbs.
$H_9 = 7500 . (45 - 2.5) - 75 . (45 - 2.5)^2 - 1406.25 = 181,875$ lbs.
$H_{10} = 7500 . (50 - 2.5) - 75 . (50 - 2.5)^2 - 1406.25 = 185,625$ lbs.

SIMPLE TRUSS. (*Fig.* 222.)

Strains in Verticals. (V_n.)

$$V_n = \frac{W}{4} - \frac{W}{2l} \cdot x_n + \frac{W_1}{4l^2} \cdot (l - x_n)^2 = 12500 - 250 . x_n +$$
$$2.5 . (l - x_n)^2$$

$V_1 = 12500 - 250 . 0 + 2.5 . 100^2 = 37,250$ lbs. Com. in U.
$V_2 = 12500 - 250 . 10 + 2.5 . 90^2 = 30,250$ lbs.
$V_3 = 12500 - 250 . 20 + 2.5 . 80^2 = 22,500$ lbs.
$V_4 = 12500 - 250 . 30 + 2.5 . 70^2 = 17,250$ lbs.
$V_5 = 12500 - 250 . 40 + 2.8 . 60^2 = 11,500$ lbs.

Counter Strains. (V_m.)

$V_6 = 12500 - 250 . 50 + 2.5 . 50^2 = 6,250$ lbs.
$V_7 = 12500 - 250 . 60 + 2.5 . 40^2 = 1,500$ lbs.

Strains in Diagonals.

$$Y_n = V_n \text{ cosec.}$$

$Y_1 = 37250 . 1.12 = 41,720$ lbs. Tension in Y_1 and Y_{10}; compression in Y_a and Y_a.

$Y_2 = 30250 . 1.12 = 33,880$ lbs. Tension in Y_2 and Y_9; compression in Y_b and Y_b.

$Y_3 = 22500 \, . \, 1.12 = 25{,}200$ lbs. Tension in Y_3 and Y_8; compression in Y_c and Y_c.

$Y_4 = 17250 \, . \, 1.12 = 19{,}320$ lbs. Tension in Y_4 and Y_7; compression in Y_d and Y_d.

$Y_5 = 11500 \, . \, 1.12 = 12{,}880$ lbs. Tension in Y_5 and Y_6; compression in Y_e and Y_e.

Counter Strains.

$$Y_m = V_m \text{ cosec. } v.$$

$Y_6 = 6250 \, . \, 1.12 = 7{,}000$ lbs. Compression in Y_5 and Y_6; tension in Y_e and Y_e.

$Y_7 = 1500 \, . \, 1.12 = 1{,}680$ lbs. Compression in Y_4 and Y_7; tension in Y_d and Y_d.

SIMPLE TRUSS. (*Fig.* 223.)

Strains in Verticals. (V_n.)

$$V_1 = 12500 - 250 \, . \, 5 + 2.5 \, . \, 95^2 = 33812.5.$$
$$V_2 = 12500 - 250 \, . \, 15 + 2.5 \, . \, 85^2 = 26812.5.$$
$$V_3 = 12500 - 250 \, . \, 25 + 2.5 \, . \, 75^2 = 20312.5.$$
$$V_4 = 12500 - 250 \, . \, 35 + 2.5 \, . \, 65^2 = 14312.5.$$
$$V_5 = 12500 - 250 \, . \, 45 + 2.5 \, . \, 55^2 = 8812.5.$$

Counter Strains. (V_m.)

$$V_6 = 12500 - 250 \, . \, 55 + 2.5 \, . \, 45^2 = 3812.$$

Strains in Diagonals.

$$Y_n = V_n \text{ cosec. } v.$$

$Y_1 = 33812.5 \, . \, 1.12 = 37{,}870$ lbs. Compression in Y_1 and Y_{10}, tension in Y_a and Y_a.

$Y_2 = 26812.5 \, . \, 1.12 = 30{,}030$ lbs. Compression in Y_2 and Y_9; tension in Y_b and Y_b.

$Y_3 = 20312.5 \, . \, 1.12 = 22{,}750$ lbs. Compression in Y_3 and Y_8; tension in Y_c and Y_c.

$Y_4 = 14312.5 \, . \, 1.12 = 16{,}030$ lbs. Compression in Y_4 and Y_7; tension in Y_d and Y_d.

$Y_5 = 8812.5 \, . \, 1.12 = 9{,}870$ lbs. Compression in Y_5 and Y_6; tension in Y_e and Y_e.

Counter Strains.

$$Y_m = V_m \text{ cosec. } v.$$

$Y_6 = 3812.5 \, . \, 1.12 = 4{,}270$ lbs. Tension in Y_5 and Y_6; compression in Y_e and Y_e.

Fig. 225.
Lower boom loaded.

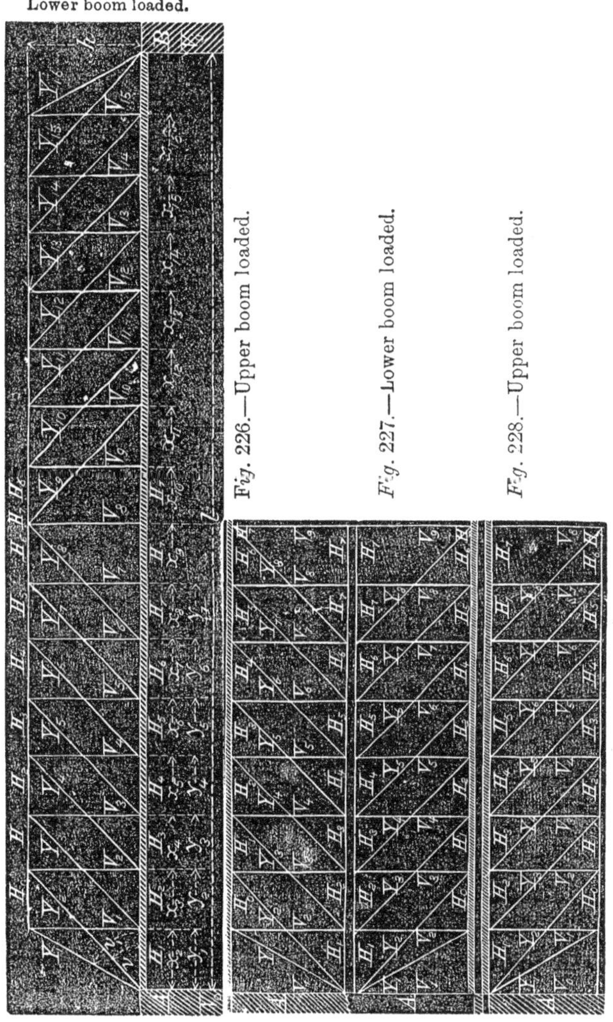

Fig. 226.—Upper boom loaded.

Fig. 227.—Lower boom loaded.

Fig. 228.—Upper boom loaded.

WHIPPLE TRUSS. (*Figs.* 225, 226, 227, and 228.)

Additional Reference.

x_n, y_n = Distance from abutment A to end of bay.

$x_1 = 0$

Static or Permanent Load, equally distributed over whole length of Truss.

Strains in Booms.

$$H_n = \frac{W}{2h} \cdot y_n - \frac{W}{2hl} \cdot y_n^2 + \frac{sW}{2hl} \cdot y_n - \frac{sW}{4h}$$

Strains in Verticals.

$$V_n = \frac{W}{4} - \frac{W}{2l} \cdot x_n$$

Strains in Diagonals.

$$Y_n = V_n \text{ cosec. } v.$$

Moving and Static Load, each equally distributed per unit of length.

Strains in Booms.

$$H_n = \frac{W + W_1}{2h} \cdot y_n - \frac{W + W_1}{2hl} \cdot y_n^2 + \frac{s(W + W_1)}{2hl} \cdot y_n - \frac{s(W + W_1)}{4h}$$

Strains in Verticals.

$$V_n = \frac{W}{4} - \frac{W}{2l} \cdot x_n + \frac{W_1}{4l^2} \cdot (l - x_n)^2$$

Strains in Diagonals.

$$Y_n = V_n \text{ cosec. } v.$$

Strains in Counters.

$$V_m = \frac{W}{4} - \frac{W}{2l} \cdot x_m + \frac{W_1}{4l^2} \cdot (l - x_m)^2 \qquad Y_m = V_m \text{ cosec. } v.$$

EXAMPLE. (*Figs.* 225, 226, 227, and 228.)

(With 20 Bays.)

Moving Load, (as railway train passing over bridge.)

Let $W = 50{,}000$ lbs.
$W_1 = 100{,}000$ lbs.
$l = 100$ feet.
$h = 10$ feet, $s = 5$ feet.
$v = 45^\circ$. (End diagonals $v = 26^\circ\ 30'$.)

Horizontal Strains in Booms. (Compression in upper, tension in lower.)

$$H_n = \frac{W + W_1}{2h} . y_n - \frac{W + W_1}{2hl} . y_n{}^2 + \frac{s(W + W_1)}{2hl} . y_n -$$

$$\frac{s(W + W_1)}{4h} = 7500 . y_n - 75 . y_n{}^2 - 375 . y_n + 18750$$

$$H_0 = 7500 \ .\ 0 - 75 \ .\ 0^2 - 375 \ .\ 0 + 18750 = 18{,}750 \text{ lbs.}$$
$$H_1 = 7500 \ .\ 5 - 75 \ .\ 5^2 - 375 \ .\ 5 + 18750 = 52{,}500 \text{ lbs.}$$
$$H_2 = 7500 \ .\ 10 - 75 \ .\ 10^2 - 375 \ .\ 10 + 18750 = 82{,}500 \text{ lbs.}$$
$$H_3 = 7500 \ .\ 15 - 75 \ .\ 15^2 - 375 \ .\ 15 + 18750 = 108{,}750 \text{ lbs.}$$
$$H_4 = 7500 \ .\ 20 - 75 \ .\ 20^2 - 375 \ .\ 20 + 18750 = 131{,}250 \text{ lbs.}$$
$$H_5 = 7500 \ .\ 25 - 75 \ .\ 25^2 - 375 \ .\ 25 + 18750 = 150{,}000 \text{ lbs.}$$
$$H_6 = 7500 \ .\ 30 - 75 \ .\ 30^2 - 375 \ .\ 30 + 18750 = 165{,}000 \text{ lbs.}$$
$$H_7 = 7500 \ .\ 35 - 75 \ .\ 35^2 - 375 \ .\ 35 + 18750 = 176{,}250 \text{ lbs.}$$
$$H_8 = 7500 \ .\ 40 - 75 \ .\ 40^2 - 375 \ .\ 40 + 18750 = 183{,}750 \text{ lbs.}$$
$$H_9 = 7500 \ .\ 45 - 75 \ .\ 45^2 - 375 \ .\ 45 + 18750 = 187{,}500$$
$$= \left(\frac{(W + W_1)\, l}{8h} \right) \text{ lbs.}$$

Strains in Verticals.

$$V_n = \frac{W}{4} - \frac{W}{2l} . x_n + \frac{W_1}{4l^2} . (l - x_n)^2 = 12500 - 250 . x_n + 2.5 . (l - x_n)^2$$

$$V_0 = \frac{W + W_1}{2} = 75{,}000 \text{ lbs.}$$

	Strains in Figs.	225	226	227	228
$V_1 = 12500 - 250 \ .\ 0 + 2.5 \ .\ 100^2$	$= 37{,}500$ lbs.	C.	C.	T.	T.
$V_2 = 12500 - 250 \ .\ 5 + 2.5 \ .\ 95^2$	$= 33{,}812$ lbs.	"	"	"	"
$V_3 = 12500 - 250 \ .\ 10 + 2.5 \ .\ 90^2$	$= 30{,}250$ lbs.	"	"	"	"
$V_4 = 12500 - 250 \ .\ 15 + 2.5 \ .\ 85^2$	$= 26{,}812$ lbs.	"	"	"	"
$V_5 = 12500 - 250 \ .\ 20 + 2.5 \ .\ 80^2$	$= 23{,}500$ lbs.	"	"	"	"
$V_6 = 12500 - 250 \ .\ 25 + 2.2 \ .\ 75^2$	$= 20{,}312$ lbs.	"	"	"	"
$V_7 = 12500 - 250 \ .\ 30 + 2.5 \ .\ 70^2$	$= 17{,}250$ lbs.	"	"	"	"

	Strains in Figs. 225	226	227	228
$V_8 = 12500 - 250 \,.\, 35 + 2.5 \,.\, 65^2 = 14{,}312$ lbs.	C.	C.	T.	T.
$V_9 = 12500 - 250 \,.\, 40 + 2.5 \,.\, 60^2 = 11{,}500$ lbs.	"	"	"	"
$V_{10} = 12500 - 250 \,.\, 45 + 2.5 \,.\, 55^2 = 8{,}812$ lbs.	"	"	"	"

V_m Acting on Counters.

$$V_{11} = 12500 - 250 \,.\, 50 + 2.5 \,.\, 50^2 = 6{,}250 \text{ lbs.}$$
$$V_{12} = 12500 - 250 \,.\, 55 + 2.5 \,.\, 45^2 = 3{,}812 \text{ lbs.}$$
$$V_{13} = 12500 - 250 \,.\, 60 + 2.5 \,.\, 40^2 = 1{,}500 \text{ lbs.}$$

Strains in Diagonals.

$$Y_n = V_n \text{ cosec. } v.$$

	Strains in Figs. 225	226	227	228
$Y_1 = 37500 \,.\, 1.117 = 41{,}887$ lbs.	Ten.	Ten.	Com.	Com.
$Y_2 = 33812 \,.\, 1.414 = 47{,}810$ lbs.	"	"	"	"
$Y_3 = 30250 \,.\, 1.414 = 42{,}773$ lbs.	"	"	"	"
$Y_4 = 26812 \,.\, 1.414 = 37{,}913$ lbs.	"	"	"	"
$Y_5 = 23500 \,.\, 1.414 = 33{,}229$ lbs.	"	"	"	"
$Y_6 = 20312 \,.\, 1.414 = 28{,}722$ lbs.	"	"	"	"
$Y_7 = 17250 \,.\, 1.414 = 24{,}391$ lbs.	"	"	"	"
$Y_8 = 14312 \,.\, 1.414 = 20{,}238$ lbs.	"	"	"	"
$Y_9 = 11500 \,.\, 1.414 = 16{,}261$ lbs.	"	"	"	"
$Y_{10} = 8812 \,.\, 1.414 = 12{,}461$ lbs.	"	"	"	"

Strains in Counters.

$$Y_{11} = 6250 \,.\, 1.414 = 8{,}837 \text{ lbs.}$$
$$Y_{12} = 3812 \,.\, 1.414 = 5{,}391 \text{ lbs.}$$
$$Y_{13} = 1500 \,.\, 1.414 = 2{,}121 \text{ lbs.}$$

[NOTE.—If counter braces are not inserted, V_{11}, V_{12}, and V_{13}, and Y_8, Y_9, and Y_{10} will have an additional strain, opposite in kind and equal to V_{11}, V_{12}, and V_{13}, and Y_{11}. Y_{12}, and Y_{13}; but if counters are used, the strain V_{11}, V_{12}, and V_{13} will not occur in the structure, but will be necessary to determine the strain in Y_{11}, Y_{12}, and Y_{13} only. Y_{11}, Y_{12}, and Y_{13} will then be inclined in the same direction as the diagonals from abutment A to center of truss, the character of strain being the same. (See also "*Howe Truss.*")

Keep in mind that each half truss, as to the character and amount of strain in the respective members, is alike.]

STRAINS IN PARABOLIC CURVED TRUSSES—"BOW-STRING GIRDERS."

(*Figs.* 229, 230, 231, 232, 233, and 234.)

The strains in the lower boom (when horizontal) are the greatest, and equal in every bay, when the load is equally distributed over the whole length.

The strains in the arch or upper boom are also greatest when the load is equally distributed over the whole length; the strains gradually increasing from the middle to the supports.

The strains in the diagonals, whether single or double, in a bay are, when the load is equally distributed, everywhere *null*. When the load is unequally distributed, and one diagonal to each bay is used, they will be either in compression or tension. The character of the maximum of strains will be as follows: Assume the left half of truss to be loaded. All diagonals inclined *up* from left to right abutment are in tension; if inclined *down*, in compression. The character of strains will be *vice versa* when the right half only is loaded.

The strains in verticals are either compression, tension, or null. The maximum of compressive strain occurs when the diagonals in connection are under the greatest strain; that is, under an unequally distributed load. For other explanation, see diagram under variously-disposed loads.

In the following formulas and examples the diagonals (for a moving load) resist a tensional strain only, and the verticals a compressive. This would not be the case if one diagonal to each bay were used. In the latter case the diagonals and verticals would have to resist an alternate compressive and tensional strain.

When the trusses are inverted, the strains are different in kind, but not in amount.

Reference.

A, B = Reaction of support.
C = Compression in arch or upper boom.
T = Tension in lower boom.
D and H = Rise of arch.
F and f = Vertical forces.
W = Weight of moving and static load per unit of span or length.
V = Strain in verticals.
N = Total number of bays.
a = Length of a bay.
c = Length of a diagonal.
d and h = Ordinates to parabola.
l = Distance between supports or span.
k = Total number of verticals = $N - 1$.
m = Number of bays between support and V_n.

n = Number of a member, counting from support to middle of truss.
t = Tension in diagonal.
v and z = Angle between horizontal and member of polygon.
w = Weight of static load per unit of span or length.
$w_{\prime}$ = Weight of moving load, equally distributed per unit of span or length.
u, x, y = Abscissas.

In the following diagrams, one-half of truss only is shown, the strains being alike in the respective members of each half:

Middle.

Fig. 229.

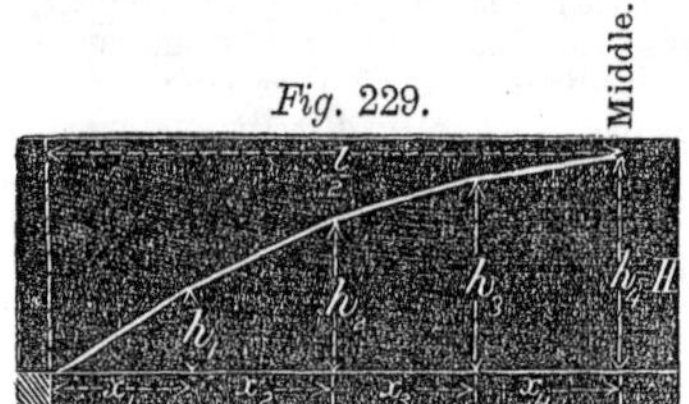

Lower Boom Horizontal.

To find the ordinates h when H is given:

$$h_n = \frac{4Hx_n(l - x_n)}{l^2}$$

The value of T given, to find h:

$$h_n = \frac{W(l - a)x_n}{2T} - \tfrac{1}{2}x_n^2 \frac{w}{T}$$

Middle.

Fig. 230.

Lower Boom Curved.

To find the ordinates h or d when H or D is given:

$$h_n = \frac{4Hx_n(l - x_n)}{l^2} \qquad d_n = \frac{4Dx_n(l - x_n)}{l^2}$$

The value of T given, to find h:

$$h_n = \frac{W(l-a)x_n}{2T} - \tfrac{1}{2}x_n{}^2\frac{w}{T}$$

Load equally distributed—Static Load. (*Figs.* 231 and 232.)
W = The weight of construction and applied load.

Fig. 231.

Lower Boom Loaded.

$$C_n = \tfrac{1}{8}\frac{Wl^2}{H}\sec. v_n \qquad T = \tfrac{1}{8}\frac{Wl^2}{H} = C \qquad V = \frac{wl}{N} = \text{tension.}$$

Upper Boom Loaded.

$$C'_n = \tfrac{1}{8}\frac{Wl^2}{H}\sec. v_n \qquad T = \tfrac{1}{8}\frac{Wl^2}{H} = C \qquad V = \text{null.}$$

Fig. 232.

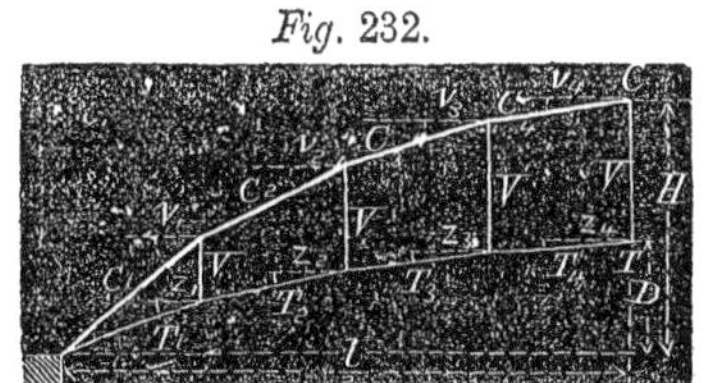

Upper Boom Loaded. ($C = T$.)

$$C_n = \tfrac{1}{8}\frac{Wl^2}{H-D}\sec. v_n \qquad T_n = \tfrac{1}{8}\frac{Wl^2}{H-D}\sec. z_n$$

$$V = \frac{lw}{N} = \text{tension.}$$

Load unequally distributed—Moving Load. (*Figs.* 233 and 234.)

(Strains in Booms, same as for Static Load.)

Fig. 233.

$$y_n = h_n \cot. v_n;\ u_n = y_n - ma$$

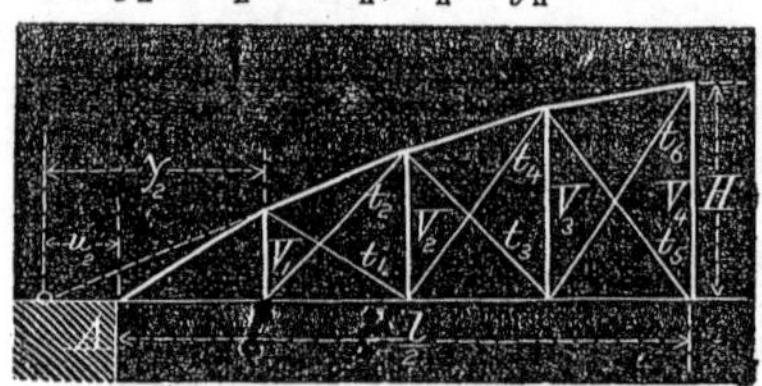

Lower Boom Loaded.

$$t_n = \frac{w_{,}l}{8H} c_n \qquad V_n = F_n - f_n = \text{compression.}$$

$$F_n = B_n \left(\frac{u_n + Na}{u_n + ma} \right) \qquad f_n = A_n \left(\frac{u_n}{u_n + ma} \right)$$

$$A_n = aw \left[\frac{(1+k-m)(k-m)}{2.N} \right] \qquad B_n = a(w + w_{,}) \left[\frac{(1+m)m}{2N} \right]$$

Upper Boom Loaded.

$$V_n = \frac{Wl}{8} = \text{compression.} \qquad t_n = \frac{w_{,}l}{8H} c_n$$

Fig. 234.

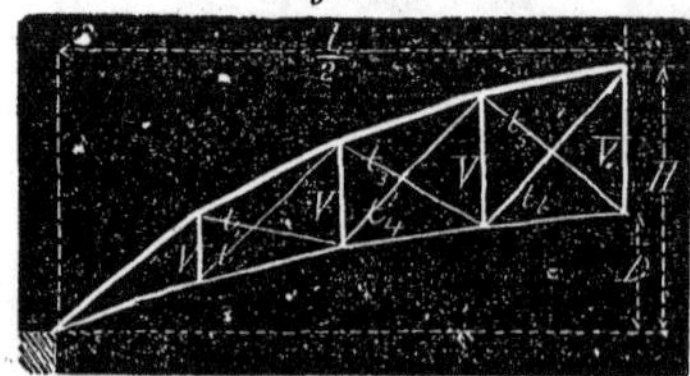

Upper Boom Loaded.

$$V_n = \frac{Wl}{8} = \text{compression.} \qquad t_n = \frac{w_{,}l}{8(H-D)} c_n$$

EXAMPLE. (*Fig.* 233.)

Moving Load on Lower Boom.

Reference.

$l = 64$ feet. $c_1 = 8.7$ feet. $w = 125$ lbs.
$H = 8$ feet. $c_2 = c_3 = 10.0$ feet. $w_{\prime} = 625$ lbs.
$a = 8$ feet. $c_4 = c_5 = 10.9$ feet. $W = w + w_{\prime} = 750$ lbs.
$N = 8,\ k = 7.$ $c_6 = 11.3$ feet.

$$h_1 = \frac{4 \times 8 \times 8(64-8)}{64^2} = 3.5 \text{ feet.}$$

$u_1 = 8.0 - 8 = 0$ feet.
$u_2 = 19.2 - 16 = 3.2$ feet.
$u_3 = 40.0 - 24 = 16.0$ feet.
$u_4 = 128.0 - 32 = 96.0$ feet.

$$h_2 = \frac{4 \times 8 \times 16(64-16)}{64^2} = 6.0 \text{ feet.}$$

$$h_3 = \frac{4 \times 8 \times 24(64-24)}{64^2} = 7.5 \text{ feet.}$$

$h_4 = H = 8.0$ feet.

$$\text{Tang. } v_1 = \frac{h_1}{a} = \frac{3.5}{8} = 23^\circ\ 37'.$$

$$\text{Tang. } v_2 = \frac{h_2 - h_1}{a} = \frac{6-3.5}{8} = 17^\circ\ 21'.$$

$$\text{Tang. } v_3 = \frac{h_3 - h_2}{a} = \frac{7.5-6}{8} = 10^\circ\ 38'.$$

$$\text{Tang. } v_4 = \frac{h_4 - h_3}{a} = \frac{8-7.5}{8} = 3^\circ\ 34'\ 30''.$$

$y_1 = 3.5 \times 2.28 = 8.0$ feet. $y_3 = 7.5 \times 5.37 = 40.0$ feet.
$y_2 = 6.0 \times 3.20 = 19.2$ feet. $y_4 = 8.0 \times 16.00 = 128.0$ feet.

$$T = C = \tfrac{1}{8}\frac{Wl^2}{H} = \tfrac{1}{8}\frac{750 \times 64^2}{8} = 48{,}000 \text{ lbs.}$$

$$C_n = C \sec.\ v_n.$$

$C_1 = 48000 \times 1.090 = 52{,}320$ lbs. $C_3 = 48000 \times 1.017 = 48{,}816$ lbs.
$C_2 = 48000 \times 1.047 = 50{,}256$ lbs. $C_4 = 48000 \times 1.0019 = 48{,}091$ lbs.

$$t_1 = \frac{625 \times 64}{8 \times 8} \times 8.7 = 5437.5 \text{ lbs.}$$

$$t_2 = t_3 = \frac{625 \times 64}{8 \times 8} \times 10.0 = 6250.0 \text{ lbs.}$$

$$t_4 = t_5 = \frac{625 \times 64}{8 \times 8} \times 10.9 = 6802.5 \text{ lbs.}$$

$$t_6 = \frac{625 \times 64}{8 \times 8} \times 11.3 = 7062.5 \text{ lbs.}$$

$$A_1 = 8 \times 125 \left[\frac{(1 + 7 - 1)(7 - 1)}{2 \times 8} \right] = 2625$$

$$A_2 = 8 \times 125 \left[\frac{(1 + 7 - 2)(7 - 2)}{2 \times 8} \right] = 1875$$

$$A_3 = 8 \times 125 \left[\frac{(1 + 7 - 3)(7 - 3)}{2 \times 8} \right] = 1250$$

$$A_4 = 8 \times 125 \left[\frac{(1 + 7 - 4)(7 - 4)}{2 \times 8} \right] = 750$$

$$B_1 = 8(125 + 625) \left[\frac{(1 + 1)1}{2 \times 8} \right] = 750$$

$$B_2 = 8(125 + 625) \left[\frac{(1 + 2)2}{2 \times 8} \right] = 2250$$

$$B_3 = 8(125 + 625) \left[\frac{(1 + 3)3}{2 \times 8} \right] = 4500$$

$$B_4 = 8(125 + 625) \left[\frac{(1 + 4)4}{2 \times 8} \right] = 7500$$

$$F_1 = 750 \left(\frac{0 + 8 \times 8}{0 + 1 \times 8} \right) = 6000.0$$

$$F_2 = 2250 \left(\frac{3.2 + 8 \times 8}{3.2 + 2 \times 8} \right) = 7812.5$$

$$F_3 = 4500 \left(\frac{16 + 8 \times 8}{16 + 3 \times 8} \right) = 9000.0$$

$$F_4 = 7500 \left(\frac{96 + 8 \times 8}{96 + 4 \times 8} \right) = 9375.0$$

$$f_1 = 2625 \left(\frac{0}{0 + 1 \times 8} \right) = 0$$

$$f_2 = 1875 \left(\frac{3.2}{3.2 + 2 \times 8} \right) = 312.5$$

$$f_3 = 1250\left(\frac{16}{16 + 3 \times 8}\right) = 500.0$$

$$f_4 = 750\left(\frac{96}{96 + 4 \times 8}\right) = 562.5$$

$V_1 = 6000 \quad - 0 \quad = 6{,}000$ lbs. $V_3 = 9000 - 500 \quad = 8{,}500$ lbs.
$V_2 = 7812.5 - 312.5 = 7{,}500$ lbs. $V_4 = 9375 - 562.5 = 8{,}812.5$ lbs.

CAPACITY AND STRENGTH OF PARABOLIC ARCHED BEAMS OR RIBS ORIGINALLY CURVED.

Reference. (*All dimensions in inches.*)

A = Sectional area of beam.
C = Compressive strain in direction of arch.
E = Modulus of elasticity.
H = Horizontal thrust at abutment, or tension on tie rod.
I = Moment of inertia of cross-section of beam.
R = Resistance of material to crushing, (to be divided by factor of safety.)
W = Concentrated load at crown of arch.
a = Vertical deflection at crown.
b = Horizontal deflection at abutments.
h = Rise of arch.
$2l$ = Distance between abutments = span.
s = Distance between neutral axis and farthest edge of section.
w = Load per unit of length, equally distributed horizontally.
x = Vertical distance from crown to point of arch, intersected by y, say at 0 on diagram.
y = Horizontal distance from middle of arch to section where the amount of strain is desired.
v = Angle between horizontal and tangent to curve.

Horizontal Thrust, (*resisted either by abutments or tie rod.*)

Fig. 235. (*All dimensions to line of pressure.*)

To determine the curve or line of pressure:

$$\frac{x}{h} = \frac{y^2}{l^2} \qquad \frac{y}{l} = \sqrt{\frac{x}{h}} \qquad y = l\sqrt{\frac{x}{h}} \qquad x = h\frac{y^2}{l^2}$$

$$\text{Tang. } v \text{ at any point} = \frac{2x}{y} = \frac{2\sqrt{hx}}{l}$$

$$\text{Tang. } v \text{ at abutment} = \frac{2h}{l}$$

Load concentrated at crown or middle of arch:

$$a = \frac{Wl_3}{256IE} \qquad b = 0 \qquad H = \tfrac{1}{2}W\left(\frac{25l}{32h} - \frac{h}{28l}\right)$$

$$C = \left(\frac{25l}{64h} - \frac{h}{56l} + \frac{hy}{l^2} - \frac{25hy^2}{32l^3}\right)W$$

$$R = \frac{25l}{64h}\frac{W}{A} + \frac{81\,Wls}{1600I}$$

$$A = \frac{25l \times 1600I}{64h(R\,1600I - 81\,Wls)}$$

Load equally distributed:

$$a = 0 \qquad b = 0 \qquad H = \frac{wl^2}{2h} \qquad C = \frac{wl^2}{2h} + \frac{why^2}{l^2}$$

$$R = \frac{C}{A} = \left(\frac{l^2}{2h} + \frac{hy^2}{l^2}\right)\frac{w}{A} \qquad A = \frac{\left(\frac{l^2}{2h} + \frac{hy^2}{l^2}\right)w}{R}$$

STRAINS IN A POLYGONAL FRAME IN EQUILIBRIUM.

Load equally distributed over members of Frame.

Reference.

H = Horizontal strain in units of weight at foot.
V_n = Vertical strain in units of weight at foot.
C_n = Compressive strain in units of weight in direction of member.
W_n = Load in units of weight, equally distributed over a member of the polygon.
v_n = Angle between horizontal and member.

Fig. 236.

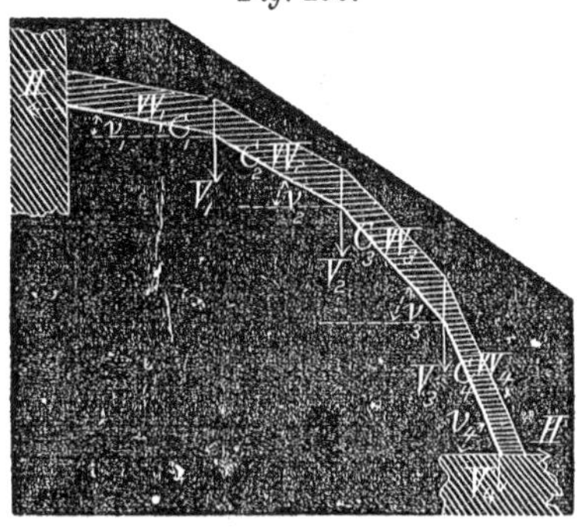

$$H = \tfrac{1}{2}\, W \text{ cotg. } v_n \qquad C_n = V_n \text{ cosec. } v_n$$

$$V_1 = \tfrac{1}{2}\, W_1$$

$$V_2 = V_1 + \frac{W_1 + W_2}{2} = \frac{W_1}{2} + \frac{W_1 + W_2}{2}$$

$$V_3 = V_2 + \frac{W_2 + W_3}{2} = \frac{W_1}{2} + \frac{W_1 + W_2}{2} + \frac{W_2 + W_3}{2}$$

$$V_4 = V_3 + \frac{W_3 + W_4}{2} = \frac{W_1}{2} + \frac{W_1 + W_2}{2} + \frac{W_2 + W_3}{2} + \frac{W_3 + W_4}{2} \;\ldots\ldots \;\&c.$$

For the equilibrium, v_1 being given:

$$\text{Tang. } v_2 = \frac{V_1}{H} = \text{tang. } v_1 + \frac{\tfrac{1}{2}(W_1 + W_2)}{H}$$

$$\text{Tang. } v_3 = \frac{V_2}{H} = \text{tang. } v_1 + \frac{\tfrac{1}{2}(W_1 + W_2) + \tfrac{1}{2}(W_2 + W_3)}{H}$$

$$\text{Tang. } v_4 = \frac{V_3}{H} = \text{tang. } v_1 + \frac{\tfrac{1}{2}(W_1 + W_2) + \tfrac{1}{2}(W_2 + W_3) + \tfrac{1}{2}(W_3 + W_4}{H}$$

The above can be used to compute the strains in ribs for dome construction.

STRAINS IN ROOF TRUSSES.

Reference. (*Figs.* 237 to 255.)

$W = \begin{cases} \text{Weight of construction.} \\ \text{Pressure of wind.} \\ \text{Pressure of snow.} \end{cases}$ = Load in units of weight, equally distributed over one rafter. (See *Fig.* 238.)

C = Compression of member in units of weight.
T = Tension of member in units of weight.
L = Total span, or distance between abutments in units of length.
d, h, l, and S = Dimensions in units of length. (See Figures.)
v, y = Angles. (See Figures.)

The diagrams show only one-half of truss, (except *Fig.* 238,) the thick lines indicating compression, and the thin ones tension. (See "*Reaction of Supports*" for pressure on joints; also "*Compound Strains in Trussed Beams*.")

Compression in Rafters. (*Trusses Nos.* 1, 3, and 4.)

The compressive strain in the rafter gradually increases from ridge to abutments. Let x = Horizontal distance from abutment to point where the strain is desired, and l half the span $= \frac{L}{2}$.

$$C \text{ for Truss No. 1} = W \sin. v \left(1 - \frac{x}{l}\right) + \frac{W}{2} \frac{\cos. v}{\text{tg.}\, v}$$

$$C \text{ for Truss No. 3} = W \sin. v \left(1 - \frac{x}{l}\right) + \frac{W}{2} \frac{\cos. v}{\text{tg.}\,(v + v_1)}$$

$$C \text{ for Truss No. 4} = W \sin. v \left(1 - \frac{x}{l}\right) + \frac{W}{2} \frac{\cos. v}{\text{tg.}\,(v - v_1)}$$

In the following examples the maximum of C is given:

Truss No. 1.

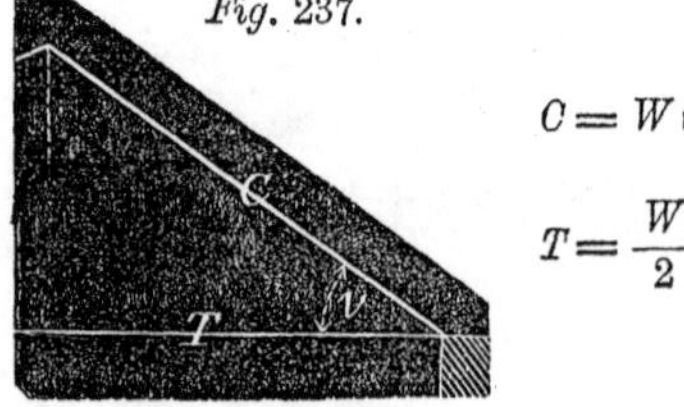

Fig. 237.

$$C = W \sin. v + \frac{W}{2} \frac{\cos. v}{\text{tg.}\, v}$$

$$T = \frac{W}{2} \text{cotg.}\, v$$

EXAMPLE.

Let $W = 8,000$ lbs.
$v = 26° 30'$.

$$C = 8000 \times 0.44619 + \frac{8000}{2} \frac{0.89493}{0.49858} = 10,666 \text{ lbs. Com.}$$

$$\text{When } x = \frac{l}{2} \text{ then will } C = \frac{8000}{2 \times 0.44619} = 8,968 \text{ lbs. Com.}$$

$$T = \frac{8000}{2} 2.00 = 8,000 \text{ lbs. Tension.}$$

Truss No. 2.

Fig. 238.

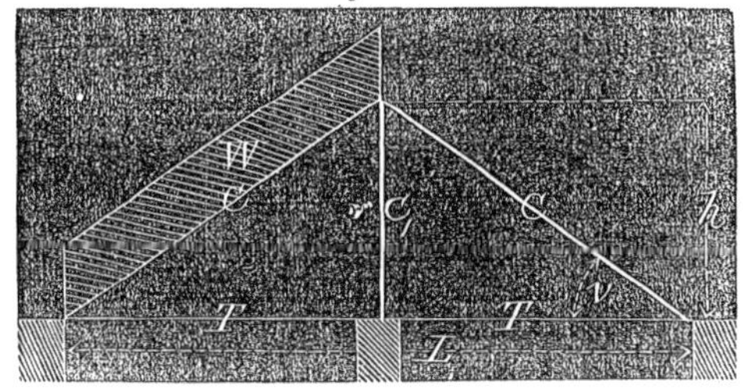

$$C = \frac{W}{2} \sin. v \qquad C_1 = W(\cos. v)^2$$

$$T = \frac{W}{2} \sin. v \cos. v = \frac{W}{4} \sin. 2v$$

EXAMPLE.

Let $W = 8,000$ lbs.
$v = 26° 30'$.

$$C = \frac{8000}{2} \times 0.4462 = 1,785 \text{ lbs. Compression.}$$

$$C_1 = 8000 \times 0.895^2 = 6,568 \text{ lbs. Compression.}$$

$$T = \frac{8000}{4} \times 0.7986 = 1,597 \text{ lbs. Tension.}$$

[NOTE.—When the rafters are fastened together at the ridge, they are under a cross-breaking strain only. Consequently there is no horizontal thrust at the abutments; that is, $T = 0$, and the compression in $C_1 = W$.]

Truss No. 3.

Fig. 239.

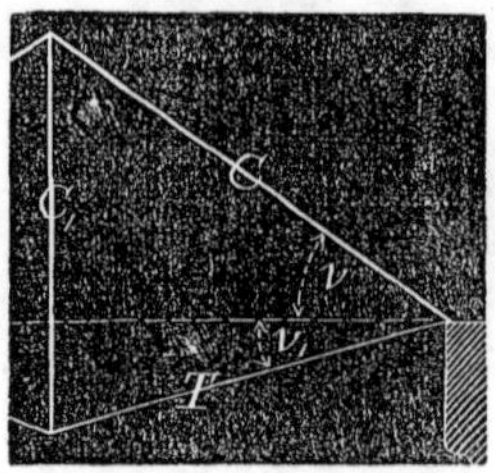

$$C = W \sin. v + \frac{W}{2} \frac{\cos. v}{\text{tg.}(v + v_1)}$$

$$C_1 = W \frac{\cos. v \sin. v_1}{\sin.(v + v_1)}$$

$$T = \frac{W}{2} \frac{\cos. v}{\sin.(v + v_1)}$$

Truss No. 4.

Fig. 240.

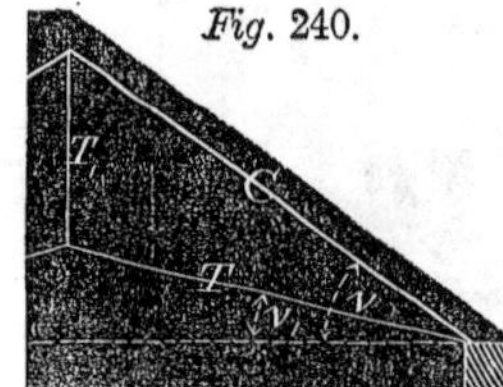

$$C = W \sin. v + \frac{W}{2} \frac{\cos. v}{\text{tg.}(v - v_1)}$$

$$T = \frac{W}{2} \frac{\cos. v}{\sin.(v - v_1)}$$

$$T_1 = W \frac{\cos. v \sin. v_1}{\sin.(v - v_1)}$$

EXAMPLE.

Let $W = 8{,}000$ lbs.
$v = 26° 30'$.
$v_1 = 5° 0'$.

$$C = 8000 \times 0.44619 + \frac{8000}{2} \frac{0.89493}{0.394} = 12{,}653 \text{ lbs. Com.}$$

$$T = \frac{8000}{2} \frac{0.894}{0.366} = 9{,}920 \text{ lbs. Tension.}$$

$$T_1 = 8000 \frac{0.894 \times 0.087}{0.366} = 1{,}720 \text{ lbs. Tension.}$$

Truss No. 5.

Fig. 241.

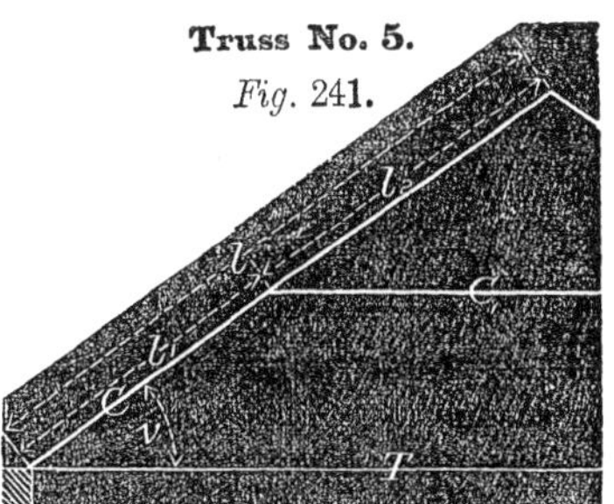

$$C = \tfrac{13}{16} W \text{ cosec. } v \quad C_1 = \tfrac{1}{2} W \text{ cotg. } v \quad T = \tfrac{1}{2}\left(1 + \frac{l_2}{l}\right) W \text{ cotg. } v$$

When there is no tie T, C_1 is under a tensile strain $= \frac{LW}{4h}$, h being the height from C_1 to ridge.

EXAMPLE.

Let $W = 8{,}000$ lbs.
$l = 22.36$ feet.
$l_1 = l_2 = = 11.18$ feet.
$v = 26° \; 30'$.

$C = \tfrac{13}{16}\, 8000 \times 2.241 = 14{,}566$ lbs. Compression.
$C_1 = \tfrac{1}{2}\, 8000 \times 2. \quad = 8{,}000$ lbs. Compression.

$$T = \tfrac{1}{2}\left(1 + \frac{11.18}{22.36}\right) 8000 \times 2. = 12{,}000 \text{ lbs. Tension.}$$

Truss No. 6.

Fig. 242.

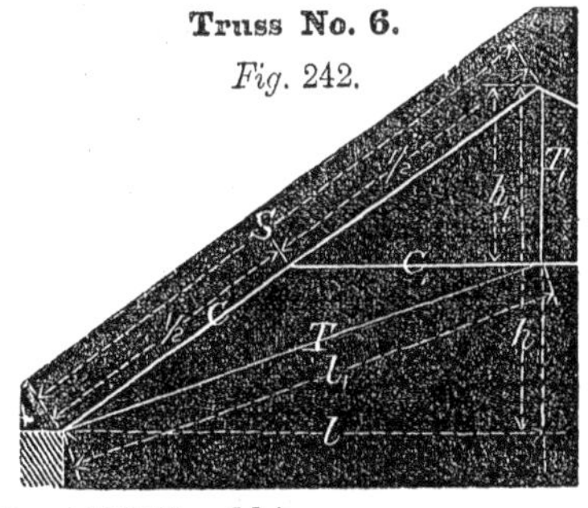

$$C = \frac{WS^2 - \tfrac{13}{16} W(S^2 - hh_1)}{h_1 S} \qquad T = \left(W - \tfrac{3}{16} W\right) \frac{l_1}{h_1}$$

$C_1 = \frac{5}{8} W \frac{l}{h}$ $\qquad$ $T_1 = 2(W - \frac{3}{16} W) \frac{h - h_1}{h_1}$

EXAMPLE.

Let $W = 8{,}000$ lbs.
$l = 20$ feet.
$l_1 = 20.6$ feet.
$h = 10$ feet.
$h_1 = 5$ feet.
$S = 22.36$.

$$C = \frac{8000 \times 500 - 1500\,(500 - 10 \times 5)}{5 \times 22.36} = 29{,}264 \text{ lbs. Com.}$$

$$C_1 = 0.625 \times 8000 \tfrac{20}{10} = 10{,}000 \text{ lbs. Compression.}$$

$$T = (8000 - 1500) \frac{20.6}{5} = 26{,}780 \text{ lbs. Tension.}$$

$$T_1 = 2(8000 - 1500) \frac{10 - 5}{5} = 13{,}000 \text{ lbs. Tension.}$$

Truss No. 7.

Fig. 243.

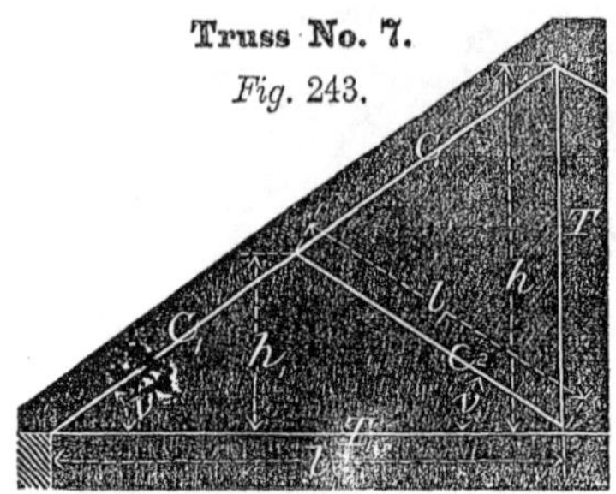

$C = W \frac{l}{2l \sin. v} = \frac{W}{2} \operatorname{cosec.} v$ $\qquad$ $C_2 = \frac{5}{8} W \frac{l_1}{h} = \frac{5}{8} W \frac{\operatorname{cosec.} v_1}{2}$

$C_1 = \frac{13}{16} W \operatorname{cosec.} v$ $\qquad$ $C_2 = \frac{5}{8} W \frac{\cos. v}{\sin. 2v}$

$T = \frac{5}{8} W \frac{h_1}{h} 2 = \frac{5}{8} W$ $\qquad$ $T_1 = \frac{13}{16} W \operatorname{cotg.} v$

EXAMPLE.

Let $W = 8{,}000$ lbs. $\quad h = 10$ feet. $\quad v = 26° 30'$.
$l = 20$ feet. $\quad l_1 = 11.18$ feet. $\quad v_1 = 26° 30'$.

$$C = 8000 \frac{20}{2 \times 20 \times 0.44619} = 8,964 \text{ lbs. Compression.}$$

$C_1 = 0.8125 \times 8000 \times 2.2411 = 14,567$ lbs. Compression.
$C_2 = 0.625 \times 8000 \times 1.12 = 5,600$ lbs. Compression.
$T = 0.625 \times 8000 = 5,000$ lbs. Tension.
$T_1 = 0.8125 \times 8000 \times 2.0 = 13,000$ lbs.

Truss No. 8.

Fig. 244.

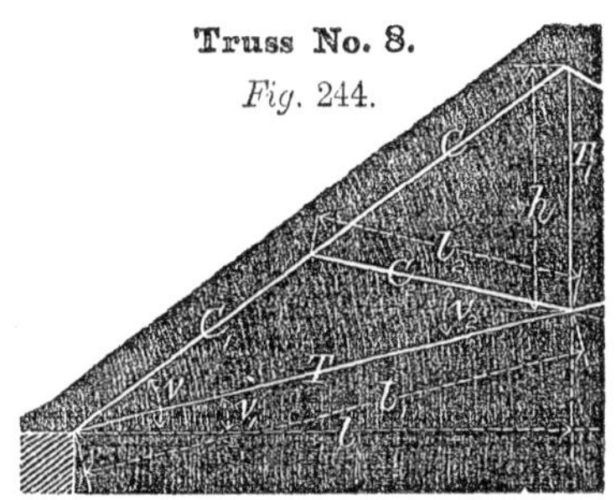

$$C = \frac{T_1 + \frac{3}{8} W}{2 \sin. v} = W \frac{l}{2 l_1 \sin.(v - v_1)} = \frac{W}{2} \frac{\cos. v_1}{\sin.(v - v_1)}$$

$$C_1 = \tfrac{13}{16} W \frac{\cos. v_1}{\sin.(v - v_1)}$$

$$C_2 = \tfrac{5}{8} W \frac{\cos. v}{\sin.(v - v_1 + v_2)} = \tfrac{5}{8} W \frac{l_2}{h}$$

$$T = \tfrac{13}{16} W \frac{\cos. v}{\sin.(v - v_1)}$$

$$T_1 = 2 W \left[\tfrac{13}{16} \frac{\cos. v \sin. v_1}{\sin.(v - v_1)} + \tfrac{5}{8} \frac{\cos. v \sin.(v_2 - v_1)}{\sin.(v - v_1 + v_2)} \right] =$$

$$2\Big(T \sin. v_1 + C_2 \sin.(v_2 - v_1)\Big) = W \frac{\sin. v \cos. v_1}{\sin.(v - v_1)} - \tfrac{3}{8} W$$

EXAMPLE.

Let $W = 8,000$ lbs.
$v = 26° 30'$.
$v_1 = 9° 20'$.
$v_2 = 19° 0'$.

$$C = \frac{9000 + 0.375 \times 8000}{0.892} = 13{,}452 \text{ lbs. Compression.}$$

$$C_1 = 0.812 \times 8000 \frac{0.986}{0.295} = 21{,}710 \text{ lbs. Compression.}$$

$$C_2 = 0.625 \times 8000 \frac{0.895}{0.590} = 7{,}585 \text{ lbs. Compression.}$$

$$T = 0.812 \times 8000 \frac{0.895}{0.295} = 19{,}702 \text{ lbs. Tension.}$$

$$T_1 = 2 \times 8000 \left[0.812 \frac{0.812 \times 0.162}{0.295} + 0.625 \times \frac{0.895 \times 0.168}{0.590} \right] = 9{,}000 \text{ lbs. Tension.}$$

Truss No. 9.

Fig. 245.

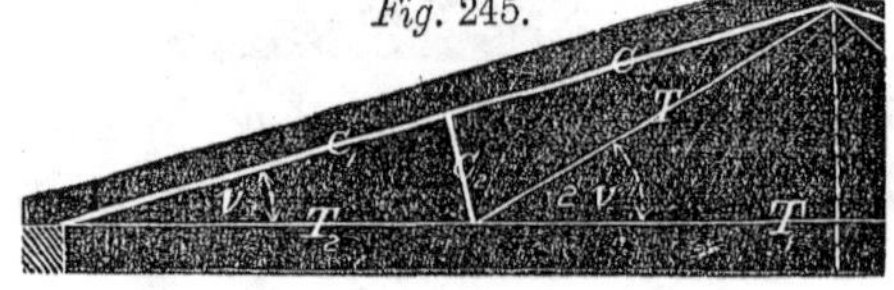

$$C = \tfrac{13}{16} W \frac{1}{\sin. v} - \tfrac{5}{8} W \sin. v$$

$$C_1 = \tfrac{13}{16} W \cdot \frac{1}{\sin. v} = \tfrac{13}{16} W \operatorname{cosec.} v$$

$C_2 = \tfrac{5}{8} W \cos. v$
$T = \tfrac{5}{16} W \operatorname{cotg.} v$
$T_1 = \tfrac{13}{16} W \operatorname{cotg.} v - \tfrac{5}{16} W \operatorname{cotg.} v = \tfrac{1}{2} W \operatorname{cotg.} v$
$T_2 = \tfrac{13}{16} W \operatorname{cotg.} v$

EXAMPLE.

Let $W = 8{,}000$ lbs.
$v = 26° \, 30'$.

$C = 0.812 \times 8000 \times 2.241 - 0.625 \times 8000 \times 0.446 = 12{,}336$ lbs. Compression.

$C_1 = 0.812 \times 8000 \times 2.241 = 14{,}566$ lbs. Compression.
$C_2 = 0.625 \times 8000 \times 0.895 = 4{,}475$ lbs. Compression.
$T = 0.312 \times 8000 \times 2. = 4{,}992$ lbs. Tension.
$T_1 = 0.812 \times 8000 \times 2 - 0.312 \times 8000 \times 2. = 8{,}000$ lbs. Tension.
$T_2 = 0.812 \times 8000 \times 2. = 12{,}992$ lbs. Tension.

Truss No. 10.

Fig. 246.

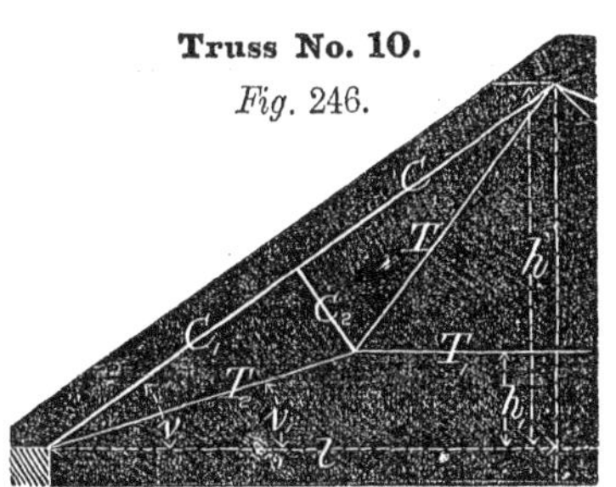

$$C = \tfrac{13}{16} W \frac{\cos.\, v_1}{\sin.(v - v_1)} - \tfrac{5}{8} W \sin.\, v$$

$$C_1 = \tfrac{13}{16} W \frac{\cos.\, v_1}{\sin.(v - v_1)}$$

$$C_2 = \tfrac{5}{8} W \cos.\, v.$$

$$T = \frac{1}{\sin.(2v - v_1)} \left[\tfrac{13}{16} W \frac{\cos.\, v \sin.\, v_1}{\sin.(v - v_1)} + \tfrac{5}{8} W \cos.^2 v \right]$$

$$T_1 = \tfrac{13}{16} W \frac{\cos.\, v \cos.\, v_1}{\sin.(v - v_1)} - T \cos.(2v - v_1) - \tfrac{5}{8} W \sin.\cos.\, v$$

$$= \frac{W}{2} \frac{l}{h - h_1}$$

$$T_2 = \tfrac{13}{16} W \frac{\cos.\, v}{\sin.(v - v_1)}$$

EXAMPLE.

Let $W = 8{,}000$ lbs. $v_1 = 9^\circ\ 20'$. $h = 10$ feet.
$v = 26^\circ\ 30'$. $l = 20$ feet. $h_1 = 2$ feet.

$$C = 0.8125 \times 8000 \frac{0.987}{0.295} - 0.625 \times 8000 \times 0.446 = 19{,}517 \text{ lbs.}$$

Compression.

$$C_1 = 0.8125 \times 8000 \frac{0.987}{0.295} = 21{,}747 \text{ lbs.}$$ Compression.

$$C_2 = 0.625 \times 8000 \times 0.895 = 4{,}475 \text{ lbs.}$$ Compression.

$$T = \frac{1}{0.6905} \left(0.8125 \times 8000 \frac{0.895 \times 0.162}{0.295} + 0.625 \times 8000 \times 0.895^2\right) = 7{,}163 \text{ lbs.}$$ Tension.

$T_1 = \frac{8000}{2} \times \frac{20}{10-2} = 10{,}000$ lbs. Tension.

$T_2 = 0.8125 \times 8000 \frac{0.895}{0.295} = 19{,}720$ lbs. Tension.

Truss No. 11.

Fig. 247.

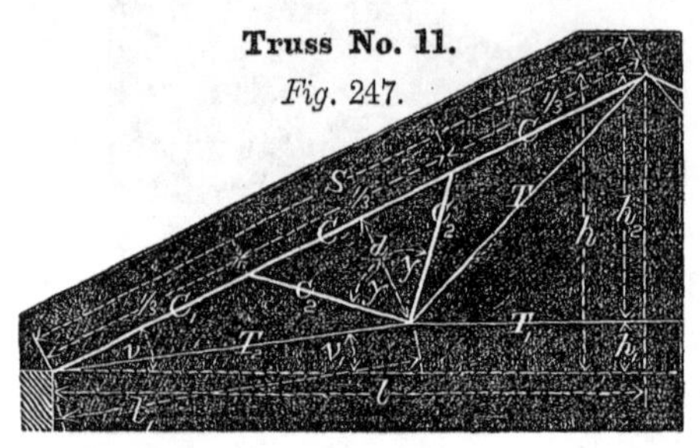

$$C = \tfrac{13}{16} W \frac{\cos. v_1}{\sin.(v - v_1)} - \tfrac{5}{8} W \sin. v$$

$$C_1 = \tfrac{13}{16} W \frac{\cos. v_1}{\sin.(v - v_1)} \qquad C_2 = \tfrac{11}{30} W \frac{\cos. v}{\cos. y}$$

$$T = \tfrac{1}{8} W \frac{\cos. v}{\sin.(2v - v_1)} \left(\tfrac{13}{2} \frac{\sin. v_1}{\sin.(v - v_1)} + 5 \cos. v\right)$$

$$T_1 = \frac{W}{2} \frac{l}{h - h_1} \qquad T_2 = \tfrac{13}{16} W \frac{\cos. v}{\sin.(v - v_1)}$$

EXAMPLE.

Let $W = 8{,}000$ lbs. $y = 50°$. $h = 10$ feet. $l = 20$ feet.
$v = 26° \; 30'$. $v_1 = 9° \; 20'$. $h_1 = 2$ feet. $S = 22.36$ feet.

$C = 0.8125 \times 8000 \frac{0.981}{0.295} - 0.625 \times 8000 \times 0.446 = 19{,}517$ lbs. Compression.

$C_1 = 0.8125 \times 8000 \frac{0.987}{0.295} = 21{,}747$ lbs. Compression.

$C_2 = 0.366 \times 8000 \frac{0.894}{0.642} = 4{,}070$ lbs. Compression.

$$T = 0.125 \times 8000 \frac{0.894}{0.690} \left(6.5 \frac{0.162}{0.295} + 5 \,.\, 0.894\right) = 11{,}050 \text{ lbs.}$$

Tension.

$$T_1 = 19486 \times 0.986 - 7421 \times 0.723 - 4930 \times 0.446 = 10{,}000 \text{ lbs.}$$

Tension.

$$T_2 = 0.812 \times 8000 \frac{0.894}{0.295} = 19{,}486 \text{ lbs.} \quad \text{Tension.}$$

Truss No. 12.

Fig. 248.

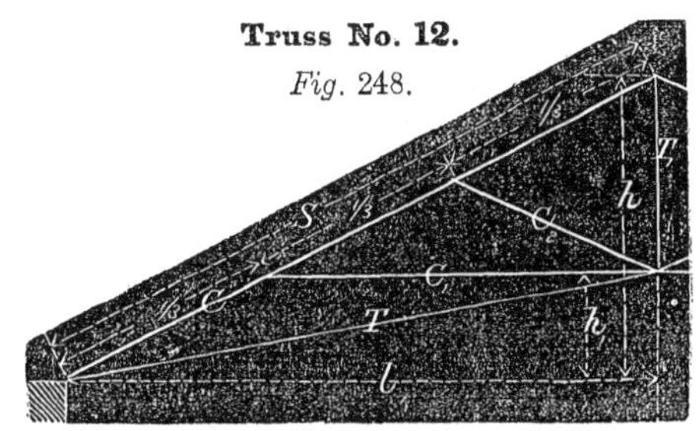

$$C = \frac{43h^2 + 39l^2}{30 \times h \times l} W \qquad T = \tfrac{13}{15} \frac{W}{2} \frac{\sqrt{h^2 + 9l^2}}{h}$$

$$C_1 = \tfrac{11}{30} W \frac{l}{h} \qquad T_1 = \tfrac{37}{30} W$$

$$C_2 = \tfrac{11}{30} \frac{W}{2} \frac{S}{h}$$

EXAMPLE.

Let $W = 8000$ lbs. $\qquad h = 10$ feet.
$l = 20$ feet. $\qquad S = 22.36$ feet.

$$C = \frac{43 \times 100 \times 15600}{30 \times 10 \times 22.36} 8000 = 23{,}704 \text{ lbs.} \quad \text{Compression.}$$

$$C_1 = 0.366 \times 8000 \frac{20}{10} = 5{,}856 \text{ lbs.} \quad \text{Compression.}$$

$$C_2 = 0.366 \times \frac{8000}{2} \frac{22.36}{10} = 3{,}280 \text{ lbs.} \quad \text{Compression.}$$

$$T = 0.866 \times \frac{8000}{2} \frac{\sqrt{100 + 3600}}{10} = 20{,}992 \text{ lbs.} \quad \text{Tension.}$$

$$T_1 = 1.23 \times 8000 = 9{,}840 \text{ lbs.} \quad \text{Tension.}$$

Truss No. 13.

Fig. 249.

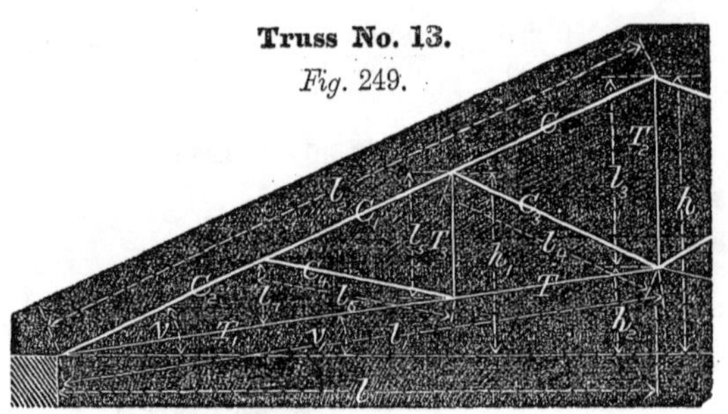

$$C = \tfrac{1}{2} W \frac{l_2}{l_3}$$

$$T = \tfrac{41}{60} W \frac{\cos. v}{\sin.(v - v_1)}$$

$$C_1 = \tfrac{41}{60} W \frac{\cos. v_1}{\sin.(v - v_1)}$$

$$T_1 = \tfrac{13}{15} W \frac{\cos. v}{\sin.(v - v_1)}$$

$$C_2 = \tfrac{13}{15} W \frac{\cos. v_1}{\sin.(v - v_1)}$$

$$T_2 = \frac{Wh}{l_3} - \tfrac{4}{15} W$$

$$C_3 = \tfrac{11}{20} W \frac{l_4}{l_3}$$

$$T_3 = \tfrac{11}{60} W$$

$$C_4 = \tfrac{11}{20} \times W \frac{l_6}{l_3}$$

EXAMPLE.

Let $W = 20{,}000$ lbs. $h = 20$ feet. $v = 21° 40$
$l = 50$ feet. $l_2 = 53.8$ feet. $v = 0°$

$$C = 0.5 \times 20000 \frac{53.8}{20} = 26{,}900 \text{ lbs. Compression.}$$

$$C_1 = 0.683 \times 20000 \frac{1}{0.369} = 37{,}018 \text{ lbs. Compression.}$$

$$C_2 = 0.866 \times 20000 \frac{1}{0.369} = 46{,}937 \text{ lbs. Compression.}$$

$$C_3 = 0.55 \times 20000 \frac{21.4}{20} = 11{,}770 \text{ lbs. Compression.}$$

$$C_4 = 0.55 \times 20000 \frac{18}{20} = 9{,}900 \text{ lbs. Compression.}$$

$T = 0.683 \times 20000 \times 2.517 = 34{,}382$ lbs. Tension.
$T_1 = 0.866 \times 20000 \times 2.517 = 43{,}594$ lbs. Tension.

$$T_2 = \frac{20000 \times 20}{20} - 5333.33 = 14{,}666 \text{ lbs. Tension.}$$

$T_3 = 0.183 \times 20000 = 3{,}660$ lbs. Tension.

Truss No. 14.

Fig. 250.

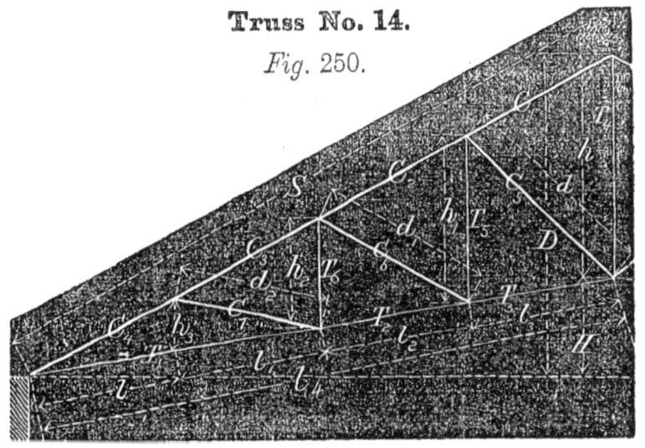

$$C_1 = \tfrac{1}{2} W \frac{S}{h} \qquad T_1 = (W - \tfrac{1}{10} W) \frac{l_4}{h}$$

$$C_2 = C_3 - \tfrac{16}{10} W \frac{S}{2h} \qquad T_2 = T_1 - \tfrac{2}{7} W \times \frac{l_1}{h_2}$$

$$C_3 = C_4 - \tfrac{2}{7} W \frac{S}{2h} \qquad T_3 = T_2 - C_6 \frac{l_2}{d_1}$$

$$C_4 = \tfrac{9}{70} W \frac{S}{h} \qquad T_4 = \frac{WD}{h} - \tfrac{1}{5} W \frac{H}{h}$$

$$C_5 = (T_5 + \tfrac{2}{7} W) \frac{d}{h} \qquad T_5 = C_6 \frac{h_2}{d_1}$$

$$C_6 = (T_6 + \tfrac{16}{70} W) \frac{d_1}{h_1} \qquad T_6 = C_7 \frac{h_3}{d_2} = \tfrac{2}{7} W \frac{h_3}{h_2}$$

$$C_7 = \tfrac{2}{7} W \frac{d_2}{h_2}$$

EXAMPLE.

Let $W = 24{,}000$ lbs. Span $= 100$ feet $l = l_1 = l_2 = l_3 = 1.25$ feet.
$h = 20$ feet. $H = 0$. $S = 53.85$ feet.

$$C_1 = 12000 \times \frac{53.85}{20} = 32{,}310 \text{ lbs. Compression.}$$

$$C_2 = 49088 - 0.228 \times 24000 \frac{53.85}{2 \times 20} = 41{,}728 \text{ lbs. Com.}$$

$$C_3 = 58320 - 0.286 \times 24000 \frac{53.85}{2 \times 20} = 49{,}088 \text{ lbs. Com.}$$

$$C_4 = 21600 \frac{53.85}{20} = 58{,}320 \text{ lbs. Compression.}$$

$$C_5 = (5801 + 0.286 \times 24000) \frac{19.5}{20} = 12{,}493 \text{ lbs. Com.}$$

$$C_6 = 3432 + 5484 \frac{16}{15} = 9{,}282 \text{ lbs. Compression.}$$

$$C_7 = 0.286 \times 24000 \frac{13.47}{10} = 9{,}245 \text{ lbs. Compression.}$$

$$T_1 = (24000 - 0.1 \times 24000) \frac{50}{20} = 54{,}000 \text{ lbs. Tension.}$$

$$T_2 = 54000 - 0.286 \times 24000 \frac{12.5}{10} = 45{,}420 \text{ lbs. Tension.}$$

$$T_3 = 45420 - 9282 \frac{12.5}{16}\ 38{,}170 \text{ lbs. Tension.}$$

$$T_4 = 24000 - \tfrac{1}{5}\, 24000 = 19{,}200 \text{ lbs. Tension.}$$

$$T_5 = 9282 \frac{10}{16} = 5{,}801 \text{ lbs. Tension.}$$

$$T_6 = 0.286 \times 24000 \frac{5}{10} = 3{,}432 \text{ lbs. Tension.}$$

Truss No. 15.

Fig. 251.

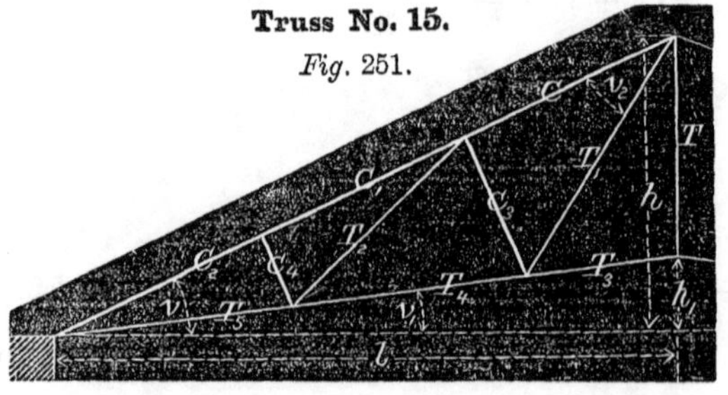

$$C = \tfrac{13}{15} W \frac{\cos. v_1}{\sin.(v - v_1)} - \tfrac{11}{15} W \sin. v - \tfrac{11}{60} W \cos. v \cot g.(v - v_1)$$

$$C_1 = \tfrac{13}{15} W \frac{\cos. v_1}{\sin.(v - v_1)} - \tfrac{11}{30} W \sin. v \qquad T_2 = \tfrac{11}{60} W \frac{\cos. v}{\sin.(v - v_1)}$$

$$C_2 = \tfrac{13}{15} W \frac{\cos. v_1}{\sin.(v - v_1)} \qquad T_3 = \frac{W}{2} \frac{l}{(h - h_1) \cos. v_1}$$

$$C_3 = \tfrac{11}{20} W \cos. v$$

$$C_4 = \tfrac{11}{30} W \cos. v \qquad T_4 = \tfrac{41}{60} W \frac{\cos. v}{\sin.(v - v_1)}$$

$$T = W \frac{l}{(h - h_1)} \operatorname{tang.} v_1 \qquad T_5 = \tfrac{13}{15} W \frac{\cos. v}{\sin.(v - v_1)}$$

$$T_1 = \frac{(T_4 - T_3) \cos.(v - v_1)}{\cos. v_2}$$

EXAMPLE.

Let $W = 20{,}000$ lbs. $h = 20$ feet. $v_1 = 0$.
$l = 50$ feet. $v = 21° 40'$. $v_2 = 46° 30'$.

$$C = 0.866 \times 20000 \frac{1}{0.369} - 0.733 \times 20000 \times 0.369 - 0.183 \times 20000 \times 0.929 \times 2.517 = 32{,}959 \text{ lbs. Compression.}$$

$$C_1 = 0.866 \times 20000 \times \frac{1}{0.369} - 0.366 \times 20000 \times 0.369 = 44{,}236 \text{ lbs. Compression.}$$

$$C_2 = 0.866 \times 20000 \times \frac{1}{0.369} = 46{,}937 \text{ lbs. Compression.}$$

$$C_3 = 0.55 \times 20000 \times 0.929 = 10{,}219 \text{ lbs. Compression.}$$

$$C_4 = 0.366 \times 20000 \times 0.929 = 6{,}800 \text{ lbs. Compression.}$$

$$T = 20000 \times \frac{40}{20} \times \operatorname{tang.} v = \text{Null.}$$

$$T_1 = \frac{(T_4 - T_3)\, 0.929}{0.688} = 10{,}920 \text{ lbs. Tension.}$$

$$T_2 = 0.183 \times 20000 \times 2.5 = 9{,}150 \text{ lbs. Tension.}$$

$$T_3 = 10000 \times \frac{50}{20 \times 1} = 25{,}000 \text{ lbs. Tension.}$$

$$T_4 = 0.683 \times 20000 \times 2.5 = 34{,}150 \text{ lbs. Tension.}$$

$$T_5 = 0.866 \times 20000 \times 2.5 = 43{,}300 \text{ lbs. Tension.}$$

Truss No. 16.

Fig. 252.

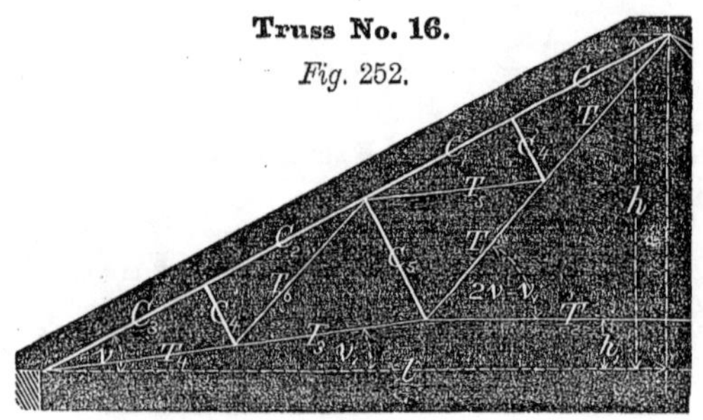

$$C = C_1 - \tfrac{2}{7} W \sin. v$$
$$C_1 = C_2 - \tfrac{16}{70} W \sin v.$$
$$C_2 = \tfrac{9}{10} W \frac{\cos. v_1}{\sin. (v - v_1)} - \tfrac{2}{7} W \sin. v$$
$$C_3 = \tfrac{9}{10} W \frac{\cos. v_1}{\sin. (v - v_1)}$$
$$C_4 = \tfrac{2}{7} W \cos. v.$$
$$C_5 = \tfrac{16}{70} W \cos. v + \tfrac{2}{7} W \cos. v = \tfrac{18}{35} W \cos. v$$
$$T = \left[\tfrac{9}{10} W \frac{\cos. v_1 \sin. v}{\sin. (v - v_1)} - \tfrac{4}{5} W \sin.^2 v - \tfrac{1}{10} W \right] \frac{1}{\sin. (2v - v_1)}$$
$$T_1 = T - \tfrac{1}{7} W \frac{\cos. v}{\sin. (v - v_1)} = T - T_5$$
$$T_2 = \frac{W}{2} \frac{l}{h - h_1}$$
$$T_3 = \tfrac{9}{10} W \frac{\cos. v}{\sin. (v - v_1)} - T_5 = \tfrac{53}{70} W \frac{\cos. v_1}{\sin. (v - v_1)}$$
$$T_4 = \tfrac{9}{10} W \frac{\cos. v}{\sin. (v - v_1)}$$
$$T_5 = T_6 = T - T_1 = \tfrac{1}{7} W \frac{\cos. v}{\sin. (v - v_1)}$$
$$T_6 = T_5$$

Example.

Let $W = 20{,}000$ lbs. $h = 20$ feet. $h_1 = 0$.
$l = 50$ feet. $v = 21°\ 40'$. $v_1 = 0$.

$C = 41885 - 0.286 \times 20000 \times 0.369 = 39{,}774$ lbs. Compression.
$C_1 = 43567 - 0.228 \times 20000 \times 0.369 = 41{,}885$ lbs. Compression.

$C_2 = 48780 - 5213 = 43{,}567$ lbs. Compression.

$C_3 = 0.9 \times 20000 \dfrac{1}{0.369} = 48{,}780$ lbs. Compression.

$C_4 = 0.286 \times 20000 \times 0.929 = 5{,}213$ lbs. Compression.
$C_5 = 0.514 \times 20000 \times 0.929 = 9{,}550$ lbs. Compression.

$$T = \left(0.9 \times 20000 \frac{0.369}{0.369} - 0.8 \times 20000 \times 0.369^2 - 0.1 \times 20000\right) \frac{1}{0.686} = 20{,}000 \text{ lbs. Tension.}$$

$T_1 = T - T_5 = 20000 - 7188 = 12{,}812$ lbs. Tension.

$T_2 = \dfrac{20000}{2} \times \dfrac{50}{20} = 25{,}000$ lbs. Tension.

$T_3 = T_4 - T_5 = 0.757 \times 20000 \dfrac{0.929}{0.369} = 38{,}118$ lbs. Tension.

$T_4 = 0.9 \times 20000 \dfrac{0.929}{0.369} = 45{,}306$ lbs. Tension.

$T_5 = T_6 = T - T_1 = 7{,}188$ lbs. Tension.
$T_6 = T_5 = 7{,}188$ lbs. Tension.

Truss No. 17.

Fig. 253.

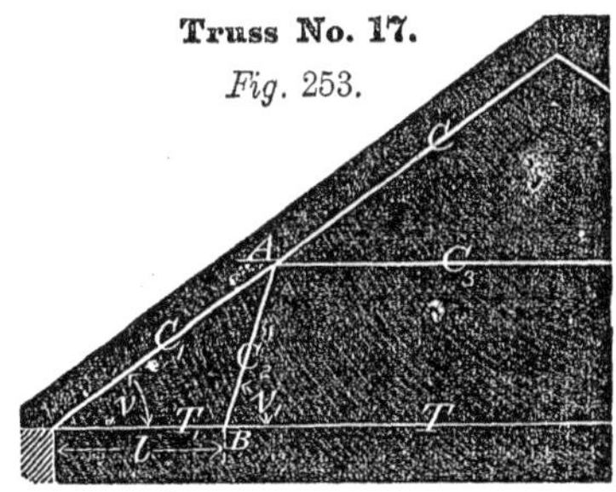

When the rafter is resting on joint A:

$C = \dfrac{W}{4 \sin. v}$ $\qquad C_3 = \frac{1}{2} \dfrac{W \cos. v \cos. (v_1 - v)}{\sin. v_1}$

$C_1 = \dfrac{W}{4 \sin. v} + \frac{1}{2} W \sin. v$ $\qquad T = C_2 \cos. v_1 + T_1$

$C_2 = \frac{1}{2} \dfrac{W \cos. v^2}{\sin. v_1}$ $\qquad T_1 = C_1 \cos. v$

Bending moment at point $B = C_2 \sin. v_1 \,.\, l$.

When rafter is fixed at joint A:

$$C = \frac{W}{4 \sin. v}$$

$$C_3 = \tfrac{1}{2} \frac{W \cos. v \cos. (v_1 - v)}{\sin. v_1}$$

$$C_1 = C$$

$$T = \tfrac{1}{2} W \text{cotg.}\, v_1 + T_1$$

$$C_2 = \tfrac{1}{2} \frac{W}{\sin. v_1}$$

$$T_1 = \frac{W}{4} \text{cotg.}\, v$$

Bending moment at $B = \frac{W}{2} . l$

Truss No. 18.

Fig. 254.

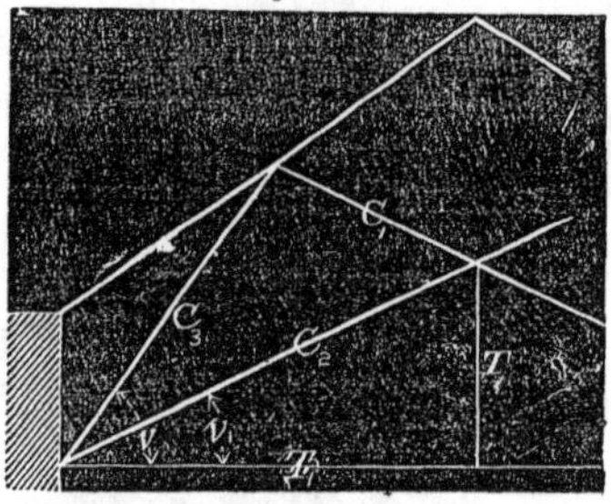

$$C_1 = \tfrac{1}{2} \frac{W \cos. v_1}{\sin. (v + v_1)}$$

$$C_2 = \tfrac{1}{2} \frac{W}{2 \sin. v_1} + C_1$$

$$C_3 = \tfrac{1}{2} \frac{W \cos. v_2}{\sin. (v + v_1)}$$

$$T = 0$$

$$T_1 = C_3 \cos. v + C_2 \cos. v_1$$

Truss No. 19.

Fig. 255.

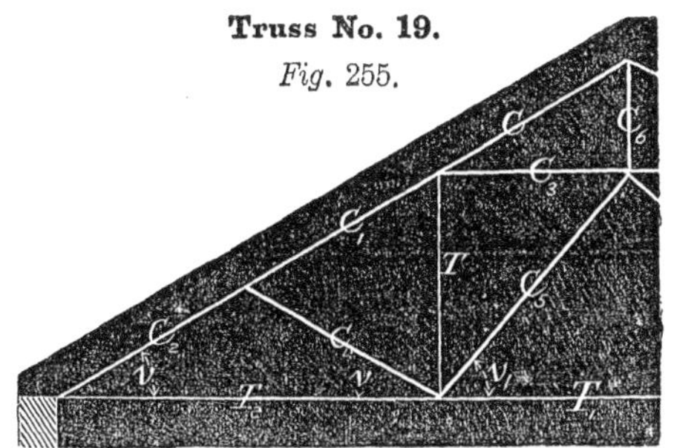

$C = \frac{1}{2} W$ cosec. v

$C_1 = \frac{41}{60} W$ cosec. v

$C_2 = \frac{13}{15} W$ cosec. v

$C_3 = \frac{2}{3} W$ cotg. v

$C_4 = \frac{1}{6} W$ cotg. v

$C_5 = \frac{1}{6} W$ tang. v_1

$C_6 = \frac{1}{3} W$

$T = \frac{1}{3} W$

$T_1 = \frac{2}{3} W$ cotg. $v + \frac{1}{6} W$ tang. v_1

$T_2 = \frac{5}{6} W$ cotg. v

EXAMPLE.

Let $W = 20{,}000$ lbs. $v = 21° 40'$. $v_1 = 56° 30'$.

$C =$	27,000 lbs.	Compression.	$C_5 =$	3,533 lbs.	Compression.
$C_1 =$	36,900 lbs.	Compression.	$C_6 =$	6,666 lbs.	Compression.
$C_2 =$	46,800 lbs.	Compression.	$T =$	6,666 lbs.	Tension.
$C_3 =$	33,466 lbs.	Compressian.	$T_1 =$	37,000 lbs.	Tension.
$C_4 =$	6,867 lbs.	Compression.	$T_2 =$	41,831 lbs.	Tension.

TABLE OF CONSTANTS.

(For Strains in Roof Trusses of those forms most in use, derived from foregoing formulas.)

Reference.

$L=$ Span in feet. $h=$ Height in feet. $C=$ Compression in member. $T=$ Tension in member.
$W=$ Weight in lbs. equally distributed over a rafter, to be multiplied by constant for strain in respective member.

$v=$ Angle between horizontal and rafter.

REFERENCE TO FIGURES.	$\frac{L}{h}=2$	$\frac{L}{h}=3$	$\frac{L}{h}=4$	$\frac{L}{h}=5$	$\frac{L}{h}=6$	$\frac{L}{h}=7$	$\frac{L}{h}=8$	$\frac{L}{h}=9$	$\frac{L}{h}=10$
	$v=45°$	$v=33° 40'$	$v=26° 30'$	$v=21° 45'$	$v=18° 20'$	$v=15° 50'$	$v=14° 15'$	$v=12° 30'$	$v=11° 10'$
Truss No. 1. (See *Fig.* 237, page 156.)	$C=1.060$ $T=0.500$	$C=1.178$ $T=0.750$	$C=1.333$ $T=1.000$	$C=1.535$ $T=1.250$	$C=1.746$ $T=1.500$	$C=1\ 969$ $T=1.750$	$C=2.154$ $T=2.000$	$C=2.417$ $T=2.250$	$C=2.678$ $T=2.500$
Truss No. 2. (See *Fig.* 238, page 157.)	$C=0.353$ $C_1=0.500$ $T=0.250$	$C=0.277$ $C_1=0.692$ $T=0.261$	$C=0.223$ $C_1=0.800$ $T=0.199$	$C=0.185$ $C_1=0.863$ $T=0.172$	$C=0.157$ $C_1=0.902$ $T=0.149$	$C=0.136$ $C_1=0.925$ $T=0.131$	$C=0.123$ $C_1=0.940$ $T=0.119$	$C=0.108$ $C_1=0.952$ $T=0.105$	$C=0.096$ $C_1=0.962$ $T=0.095$
Truss No. 5. (See *Fig.* 241, page 159.)	$C=1.456$ $C_1=0.500$ $T=0.750$	$C=1.465$ $C_1=0.750$ $T=1.125$	$C=1.820$ $C_1=1.000$ $T=1.500$	$C=2.194$ $C_1=1.250$ $T=1.875$	$C=2.576$ $C_1=1.500$ $T=2.250$	$C=2.974$ $C_1=1.750$ $T=2.625$	$C=3.315$ $C_1=2.000$ $T=3.000$	$C=3.754$ $C_1=2.250$ $T=3.375$	$C=4.193$ $C_1=2.500$ $T=3.750$

$h_1 = \frac{h}{2}$ **Truss No. 6.** (See *Fig.* 242, page 159.)	$C = 2.234$ $C_1 = 0.625$ $T = 1.625$ $T_1 = 1.625$	$C = 2.951$ $C_1 = 0.937$ $T = 2.428$ $T_1 = 1.625$	$C = 3.658$ $C_1 = 1.250$ $T = 3.259$ $T_1 = 1.625$	$C = 4.363$ $C_1 = 1.563$ $T = 4.180$ $T_1 = 1.625$	$C = 5.135$ $C_1 = 1.875$ $T = 4.770$ $T_1 = 1.625$	$C = 5.916$ $C_1 = 2.187$ $T = 5.781$ $T_1 = 1.625$	$C = 6.549$ $C_1 = 2.500$ $T = 6.460$ $T_1 = 1.625$	$C = 7.524$ $C_1 = 2.812$ $T = 7.320$ $T_1 = 1.625$	$C = 8.450$ $C_1 = 3.125$ $T = 8.188$ $T_1 = 1.625$
$v_1 = v$ **Truss No. 7.** (See *Fig.* 243.) page 160.)	$C_1 = 1.145$ $C_2 = 0.442$ $T = 0.625$ $T_1 = 0.812$	$C_1 = 1.462$ $C_2 = 0.560$ $T = 0.625$ $T_1 = 1.219$	$C_1 = 1.820$ $C_2 = 0.700$ $T = 0.625$ $T_1 = 1.625$	$C_1 = 2.194$ $C_2 = 0.844$ $T = 0.625$ $T_1 = 2.031$	$C_1 = 2.575$ $C_2 = 0.987$ $T = 0.625$ $T_1 = 2.437$	$C_1 = 2.974$ $C_2 = 1.144$ $T = 0.625$ $T_1 = 2.843$	$C_1 = 3.313$ $C_2 = 1.275$ $T = 0.625$ $T_1 = 3.250$	$C_1 = 3.754$ $C_1 = 1.444$ $T = 0.625$ $T_1 = 3.656$	$C_1 = 4.192$ $C_2 = 1.612$ $T = 0.625$ $T_1 = 4.062$
Truss No. 9. (See *Fig.* 245, page 162.)	$C_1 = 1.145$ $C_2 = 0.442$ $T = 0.312$ $T_1 = 0.500$ $T_2 = 0.812$	$C_1 = 1.462$ $C_2 = 0.520$ $T = 0.469$ $T_1 = 0.750$ $T_2 = 1.219$	$C_1 = 1.820$ $C_2 = 0.559$ $T = 0.625$ $T_1 = 1.000$ $T_2 = 1.625$	$C_1 = 2.194$ $C_2 = 0.531$ $T = 0.731$ $T_1 = 1.250$ $T_2 = 2.031$	$C_1 = 2.575$ $C_2 = 0.594$ $T = 0.937$ $T_1 = 1.500$ $T_2 = 2.445$	$C_1 = 2.974$ $C_2 = 0.600$ $T = 1.094$ $T_1 = 1.750$ $T_2 = 2.860$	$C_1 = 3.313$ $C_2 = 0.606$ $T = 1.250$ $T_1 = 2.000$ $T_2 = 3.217$	$C_1 = 3.754$ $C_2 = 0.610$ $T = 1.406$ $T_1 = 2.250$ $T_2 = 3.664$	$C_1 = 4.192$ $C_2 = 0.613$ $T = 1.562$ $T_1 = 2.500$ $T_2 = 4.111$
$h_1 = \frac{h}{3}$ **Truss No. 12.** (See *Fig.* 248, page 165.)	$C = 1.932$ $C_1 = 0.366$ $C_2 = 0.258$ $T = 1.375$ $T_1 = 1.233$	$C = 2.333$ $C_1 = 0.549$ $C_2 = 0.329$ $T = 2.000$ $T_1 = 1.233$	$C = 2.963$ $C_1 = 0.732$ $C_2 = 0.410$ $T = 2.625$ $T_1 = 1.233$	$C = 3.553$ $C_1 = 0.915$ $C_2 = 0.494$ $T = 3.250$ $T_1 = 1.233$	$C = 4.170$ $C_1 = 1.100$ $C_2 = 0.590$ $T = 3.875$ $T_1 = 1.233$	$C = 4.780$ $C_1 = 1.290$ $C_2 = 0.670$ $T = 4.500$ $T_1 = 1.233$	$C = 5.339$ $C_1 = 1.449$ $C_2 = 0.747$ $T = 5.125$ $T_1 = 1.233$	$C = 6.030$ $C_1 = 1.651$ $C_2 = 0.845$ $T = 5.750$ $T_1 = 1.233$	$C = 6.723$ $C_1 = 1.852$ $C_2 = 0.944$ $T = 6.375$ $T_1 = 1.233$

TABLE OF CONSTANTS—Continued.

REFERENCE TO FIGURES.	$\frac{L}{h}=2$	$\frac{L}{h}=3$	$\frac{L}{h}=4$	$\frac{L}{h}=5$	$\frac{L}{h}=6$	$\frac{L}{h}=7$	$\frac{L}{h}=8$	$\frac{L}{h}=9$	$\frac{L}{h}=10$
	$v=45°$	$v=33° 40'$	$v=26° 30'$	$v=21° 45'$	$v=18° 20'$	$v=15° 50'$	$v=14° 15'$	$v=12° 30'$	$v=11° 10'$
$v_1=0$	$C_2=1.211$	$C_2=1.559$	$C_2=1.940$	$C_2=2.338$	$C_2=2.745$	$C_2=3.170$	$C_2=3.533$	$C_2=4.000$	$C_2=4.468$
	$C_3=0.404$	$C_3=0.460$	$C_3=0.523$	$C_3=0.594$	$C_3=0.660$	$C_3=0.731$	$C_3=0.797$	$C_3=0.902$	$C_3=0.990$
	$C_4=0.261$	$C_4=0.330$	$C_4=0.413$	$C_4=0.495$	$C_4=0.567$	$C_4=0.661$	$C_4=0.742$	$C_4=0.837$	$C_4=0.936$
Truss No. 13.	$T'=0.683$	$T'=1.024$	$T'=1.366$	$T'=1.707$	$T'=1.956$	$T'=2.304$	$T'=2.705$	$T'=3.080$	$T'=3.456$
(See *Fig.* 249, page 166.)	$T_1=0.866$	$T_1=1.299$	$T_1=1.732$	$T_1=2.165$	$T_1=2.606$	$T_1=3.048$	$T_1=3.429$	$T_1=3.905$	$T_1=4.382$
	$T_2=0.734$	$T_2=0.734$	$T_2=0.734$	$T_2=0.734$	$T_2=0.734$	$T_2=0.734$	$T_2=0.734$	$T_2=0.734$	$T_2=0.734$
	$T_3=0.183$	$T_3=0.183$	$T_3=0.183$	$T_3=0.183$	$T_3=0.183$	$T_3=0.183$	$T_3=0.183$	$T_3=0.183$	$T_3=0.183$
	$C_4=1.269$	$C_4=1.620$	$C_4=2.016$	$C_4=2.430$	$C_4=2.853$	$C_4=3.294$	$C_4=3.672$	$C_4=4.158$	$C_4=4.644$
	$C_5=0.420$	$C_5=0.476$	$C_5=0.447$	$C_5=0.521$	$C_5=0.563$	$C_5=0.616$	$C_5=0.670$	$C_5=0.731$	$C_5=0.781$
$H=0$	$C_6=0.285$	$C_6=0.305$	$C_6=0\ 349$	$C_6=0.393$	$C_6=0.428$	$C_6=0.515$	$C_6=0.571$	$C_6=0.605$	$C_6=0.667$
$h=D$	$C_7=0.203$	$C_7=0.257$	$C_7=0.320$	$C_7=0.386$	$C_7=0.451$	$C_7=0.522$	$C_7=0.582$	$C_7=0.660$	$C_7=0.737$
	$T_1=0.900$	$T_1=1.350$	$T_1=1.800$	$T_1=2.250$	$T_1=2.710$	$T_1=3.168$	$T_1=3.560$	$T_1=4.059$	$T_1=4.554$
	$T_2=0.757$	$T_2=1.136$	$T_2=1.514$	$T_2=1.893$	$T_2=2.279$	$T_2=2.665$	$T_2=2.998$	$T_2=3.415$	$T_2=3.831$
Truss No. 14.	$T_3=0.631$	$T_3=0.921$	$T_3=1.267$	$T_3=1.589$	$T_3=1.926$	$T_3=2.222$	$T_3=2.496$	$T_3=2.862$	$T_3=3.214$
(See *Fig.* 250.) page 167.)	$T_4=0.800$	$T_4=0.800$	$T_4=0.800$	$T_4=0.800$	$T_4=0.800$	$T_4=0.800$	$T_4=0.800$	$T_4=0.800$	$T_4=0.800$
	$T_5=0.253$	$T_5=0.253$	$T_5=0.253$	$T_5=0.253$	$T_5=0.253$	$T_5=0.253$	$T_5=0.253$	$T_5=0.253$	$T_5=0.253$
	$T_6=0.143$	$T_6=0.143$	$T_6=0.143$	$T_6=0.143$	$T_6=0.143$	$T_6=0\ 143$	$T_6=0.143$	$T_6=0.143$	$T_6=0.143$

$v_1 = 0$		$C_2=1.560$	$C_2=1.940$	$C_2=2.340$	$C_2=2.745$	$C_2=3.170$	$C_2=3.533$	$C_2=4.000$	$C_2=4.170$
$T = 0$		$C_3=0.457$	$C_3=0.404$	$C_3=0.511$	$C_3=0.523$	$C_3=0.528$	$C_3=0.534$	$C_3=0.537$	$C_3=0.539$
$h_1 = 0$		$C_4=0.305$	$C_4=0.328$	$C_4=0.341$	$C_4=0.349$	$C_4=0.352$	$C_4=0.356$	$C_4=0.359$	$C_4=0.360$
		$T_1=0.376$	$T_1=0.446$	$T_1=0.546$	$T_1=0.627$	$T_1=0.711$	$T_1=0.788$	$T_1=0.885$	$T_1=1.015$
		$T_2=0.274$	$T_2=0.366$	$T_2=0.457$	$T_2=0.551$	$T_2=0.644$	$T_2=0.725$	$T_2=0.825$	$T_2=0.926$
Truss No. 15.		$T_3=0.750$	$T_3=1.000$	$T_3=1.250$	$T_3=1.500$	$T_3=1.750$	$T_3=2.000$	$T_3=2.250$	$T_3=2.500$
(See *Fig.* 251.)		$T_4=1.025$	$T_4=1.366$	$T_4=1.710$	$T_4=2.056$	$T_4=2.404$	$T_4=2.705$	$T_4=3.080$	$T_4=3.456$
page 168.)		$T_5=1.300$	$T_5=1.732$	$T_5=2.165$	$T_5=2.607$	$T_5=3.048$	$T_5=3.430$	$T_5=3.906$	$T_5=4.382$
	$C_3=1.269$	$C_3=2.620$	$C_3=2.016$	$C_3=2.430$	$C_3=3.853$	$C_3=3.294$	$C_3=3.672$	$C_3=4.158$	$C_3=4.644$
	$C_4=0.202$	$C_4=0.238$	$C_4=0.255$	$C_4=0.265$	$C_4=0.271$	$C_4=0.274$	$C_4=0.277$	$C_4=0.279$	$C_4=0.280$
$v_1 = 0$	$C_5=0.363$	$C_5=0.428$	$C_5=0.460$	$C_5=0.478$	$C_5=0.488$	$C_5=0.493$	$C_5=0.499$	$C_5=0.502$	$C_5=0.504$
$h_1 = 0$	$T=0.400$	$T=0.600$	$T=0.800$	$T=1.000$	$T=1.200$	$T=1.400$	$T=1.600$	$T=1.800$	$T=2.000$
	$T_1=0.257$	$T_1=0.386$	$T_1=0.514$	$T_1=0.643$	$T_1=0.770$	$T_1=0.897$	$T_1=1.034$	$T_1=1.155$	$T_1=1.276$
	$T_2=0.500$	$T_2=0.750$	$T_2=1.000$	$T_2=1.250$	$T_2=1.500$	$T_2=1.750$	$T_2=2.000$	$T_2=2.250$	$T_2=2.500$
Truss No. 16.	$T_3=0.700$	$T_3=1.050$	$T_3=1.400$	$T_3=1.750$	$T_3=2.100$	$T_3=2.450$	$T_3=2.800$	$T_3=3.150$	$T_3=3.500$
(See *Fig.* 252,	$T_4=0.900$	$T_4=1.350$	$T_4=1.800$	$T_4=2.250$	$T_4=2.709$	$T_4=3.168$	$T_4=3.564$	$T_4=4.059$	$T_4=4.554$
page 170.)	$T_5=0.143$	$T_5=0.214$	$T_5=0.286$	$T_5=0.357$	$T_5=0.430$	$T_5=0.503$	$T_5=0.567$	$T_5=0.645$	$T_5=0.724$
	$T_6=0.143$	$T_6=0.214$	$T_6=0.286$	$T_6=0.357$	$T_6=0.430$	$T_6=0.503$	$T_6=0.567$	$T_6=0.615$	$T_6=0.724$

EXAMPLE TO TABLE OF CONSTANTS. (*Truss No.* 13.)

What is the amount of strain in the various members of a truss, according to *Fig.* 249, of the following dimensions, viz: Span 60 feet, distance between trusses 10 feet, height at center 10 feet, weight to be carried, including weight of construction, $66\frac{2}{3}$ lbs. per square foot horizontally; hence total weight on one rafter $= 30 \times 10 \times 66\frac{2}{3} = 20{,}000$ lbs.?

$L = 60$ feet. $h = 10$ feet. $\frac{L}{h} = \frac{60}{10} = 6.$ $v = 18^\circ\ 20'.$ $W = 20{,}000$ lbs.

Member.	Constant.	W	Strains.	
C_2	$= 2.745 \times$	$20{,}000 =$	54,900 lbs.	Compression.
C_3	$= 0.660 \times$	$20{,}000 =$	13,200 lbs.	Compression.
C_4	$= 0.567 \times$	$20{,}000 =$	11,340 lbs.	Compression.
T	$= 1.956 \times$	$20{,}000 =$	39,120 lbs.	Tension.
T_1	$= 2.606 \times$	$20{,}000 =$	52,120 lbs.	Tension.
T_2	$= 0.734 \times$	$20{,}000 =$	14,680 lbs.	Tension.
T_3	$= 0.183 \times$	$20{,}000 =$	3,660 lbs.	Tension.

[NOTE.—In the foregoing table the proportion of h to L is approximate. The constants are based on the angles.]

PRESSURE OF WIND ON ROOFS.

In the following table the maximum pressure of wind is taken at 50 lbs. per square foot:

The angle between horizontal and direction of wind is generally $10^\circ\ 00'$. (See diagram.)

Fig. 256.

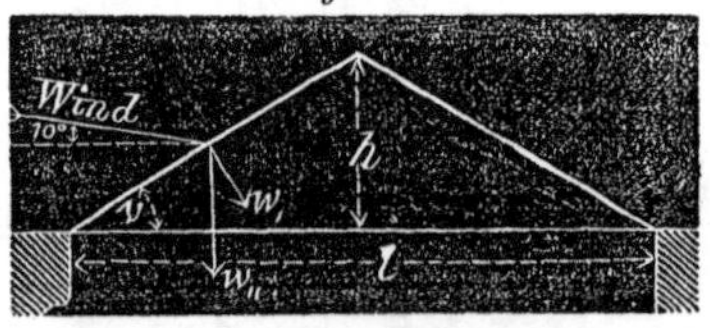

Reference.

F = Force of wind in lbs. = 50.
$w_{\prime}$ = Pressure at right angles to surface per square foot in lbs.
$w_{\prime\prime}$ = Pressure, vertical, per square foot in lbs.

$$w_{\prime} = F \sin.^2 (v + 10)$$

$$w_{\prime\prime} = \frac{w_{\prime}}{\cos.\ v}$$

Proportion of height h to span l.	Angle v.	Pressure $w_{\prime}$ in lbs.	Pressure $w_{\prime\prime}$ in lbs.
$h=\frac{l}{0}$	90° 00′	50.00	0.00
$h=\frac{l}{2}$	45° 00′	33.53	47.40
$h=\frac{l}{3}$	33° 41′ 50″	23.80	28.60
$h=\frac{l}{4}$	26° 33′ 50″	17.64	19.70
$h=\frac{l}{5}$	21° 48′	13.83	14.80
$h=\frac{l}{6}$	18° 26′	11.23	11.80
$h=\frac{l}{7}$	15° 54′ 40″	9.46	9.80
$h=\frac{l}{8}$	14° 02′ 10″	8.56	8.80
$h=\frac{l}{9}$	12° 31′ 40″	7.29	7.40
$h=\frac{l}{10}$	11° 18′ 40″	6.51	6.60

PRESSURE OF SNOW ON ROOFS.

The average pressure of snow on a level surface, in the United States, is about 15 lbs. per square foot.

The following table gives the pressure per square foot on various inclined surfaces:

Reference.

P = Pressure per square foot in lbs. = 15.
p_1 = Vertical pressure in lbs.
p_2 = Pressure at right angles to surface in lbs.
v = Angle between surface and horizontal.

$p_1 = P \cos. v.$
$p_2 = p_1 \cos. v.$

Proportion of height h to span l.	Angle v.	Pressure P_1 in lbs.	Pressure P_2 in lbs.
$h = \frac{l}{2}$	45° 00′	10.60	7.49
$h = \frac{l}{3}$	33° 41′ 50″	12.48	10.38
$h = \frac{l}{4}$	26° 33′ 50″	13.42	12.00
$h = \frac{l}{5}$	21° 48′	13.93	12.94
$h = \frac{l}{6}$	18° 26′	14.23	13.50
$h = \frac{l}{7}$	15° 54′ 40″	14.41	13.86
$h = \frac{l}{8}$	14° 02′ 10″	14.52	14.05
$h = \frac{l}{9}$	12° 31′ 40″	14.64	14.29
$h = \frac{l}{10}$	11° 18′ 40″	14.71	14.43
$h = \frac{l}{\infty}$	0° 00′ 00″	15.00	15.00

TIE RODS AND BARS.

Capacity and Proportional Dimensions of Wrought-iron Tie Rods, Tie Bars, and Pins or Bolts.

Ultimate resistance to tearing = 60,000 lbs. = 30 tons per square inch.

Ultimate resistance to shearing = 50,000 lbs. = 25 tons per square inch. (See *Fig.* 258.)

Capacity of tie or bar.				Sectional area in sq. inches.	Diameter in inches, if round.	Dimension of flat bars in in., uniform thickness.			Diameter *D* of pin or bolt.	
3 times safety.		5 times safety.				Thickness *t* of bar.	Width *b* of bar.	Width *b* around eye.	One place of shearing.	Two places of shearing.
Lbs.	Tons.	Lbs.	Tons.							
5,000	2.50	3,000	1.50	0.25	0.56	1/4	1	0.75	0.62	0.44
6,200	3.10	3,720	1.86	0.31	0.62	"	1 1/4	0.93	0.69	0.48
7,400	3.70	4,440	2.22	0.37	0.70	"	1 1/2	1.12	0.75	0.53
8,600	4.30	5,160	2.58	0.43	0.74	"	1 3/4	1.31	0.80	0.57
10,000	5.00	6,000	3.00	0.50	0.79	"	2	1.50	0.88	0.62
11,200	5.60	6,720	3.36	0.56	0.84	"	2 1/4	1.68	0.92	0.65
12,400	6.20	7,440	3.72	0.62	0.89	"	2 1/2	1.87	0.97	0.69
13,600	6.80	8,160	3.88	0.68	0.93	"	2 3/4	2.06	1.01	0.72
15,000	7.50	9,000	4.50	0.75	0.97	"	3	2.25	1.08	0.76
7,400	3.70	4,440	2.22	0.37	0.68	3/8	1	0.75	0.75	0.53
9,200	4.60	5,520	2.76	0.46	0.76	"	1 1/4	0.93	0.83	0.58
11,200	5.60	6,720	3.36	0.56	0.84	"	1 1/2	1.12	0.92	0.65
13,000	6.50	7,800	3.90	0.65	0.91	"	1 3/4	1.31	0.99	0.70
15,000	7.50	9,000	4.50	0.75	0.97	"	2	1.50	1.08	0.76
16,800	8.40	10,080	5.04	0.84	1.04	"	2 1/4	1.68	1.13	0.80
18,600	9.30	11,160	5.58	0.93	1.09	"	2 1/2	1.87	1.19	0.83
20,600	10.30	12,360	6.18	1.03	1.15	"	2 3/4	2.06	1.24	0.83
22,400	11.20	13,440	6.72	1.12	1.19	"	3	2.25	1.29	0.92
10,000	5.00	6,000	3.00	0.50	0.79	1/2	1	0.75	0.88	0.62
12,400	6.20	7,440	3.72	0.62	0.88	"	1 1/4	0.93	0.97	0.69
15,000	7.50	9,000	4.50	0.75	0.97	"	1 1/2	1.12	1.08	0.76
17,400	8.70	10,440	5.02	0.87	1.05	"	1 3/4	1.31	1.16	0.82
20,000	10.00	12,000	6.00	1.00	1.13	"	2	1.50	1.24	0.88
22,400	11.20	13,440	6.72	1.12	1.20	"	2 1/4	1.68	1.32	0.93
25,000	12.50	15,000	7.50	1.25	1.26	"	2 1/2	1.87	1.39	0.98
27,400	13.70	16,440	8.22	1.37	1.32	"	2 3/4	2.06	1.45	1.03
30,000	15.00	18,000	9.00	1.50	1.39	"	3	2.25	1.52	1.08
12,400	6.20	7,440	3.72	0.62	0.90	5/8	1	0.75	0.98	0.69
15,600	7.80	9,360	4.68	0.78	1.00	"	1 1/4	0.93	1.09	0.77
18,600	9.30	11,160	5.58	0.93	1.09	"	1 1/2	1.12	1.20	0.85
21,800	10.90	13,080	6.54	1.09	1.18	"	1 3/4	1.31	1.29	0.91
25,000	12.50	15,000	7.50	1.25	1.26	"	2	1.50	1.39	0.98
28,000	14.00	16,800	8.40	1.40	1.34	"	2 1/4	1.68	1.47	1.04
30,533	15.27	18,720	9.36	1.56	1.41	"	2 1/2	1.87	1.54	1.09

Capacity of tie or bar.				Sectional area in sq. inches.	Diameter in inches, if round.	Dimension of flat bars in in., uniform thickness.			Diameter D of pin or bolt.	
3 times safety.		5 times safety.				Thickness t of bar	Width b_1 of bar.	Width b around eye.	One place of shearing.	Two places of shearing.
Lbs.	Tons.	Lbs.	Tons.							
34,200	17.10	20,520	10.26	1.71	1.48	⅝	2¾	2.06	1.62	1.14
37,500	18.75	22,440	11.22	1.87	1.54	"	3	2.25	1.69	1.20
15,000	7.50	9,000	4.50	0.75	0.98	¾	1	0.75	1.08	0.76
18,600	9.30	11,160	5.58	0.93	1.09	"	1¼	0.93	1.20	0.85
22,400	11.20	13,440	6.72	1.12	1.19	"	1½	1.12	1.31	0.93
26,200	13.10	15,720	7.86	1.31	1.30	"	1¾	1.31	1.41	1.00
30,000	15.00	18,000	9.00	1.50	1.39	"	2	1.50	1.52	1.08
33,600	16.80	20,160	10.08	1.68	1.46	"	2¼	1.68	1.62	1.14
37,400	18.70	22,440	11.22	1.87	1.54	"	2½	1.87	1.69	1.20
41,200	20.60	24,720	12.36	2.06	1.62	"	2¾	2.06	1.77	1.26
45,000	22.50	27,000	13.50	2.25	1.69	"	3	2.25	1.86	1.32
17,400	8.70	10,440	5.22	0.87	1.05	⅞	1	0.75	1.16	0.82
21,800	10.90	13,080	6.54	1.09	1.18	"	1¼	0.93	1.29	0.91
26,200	13.10	15,720	7.86	1.31	1.29	"	1½	1.12	1.41	1.00
30,600	15.30	18,360	9.18	1.53	1.40	"	1¾	1.31	1.53	1.08
34,800	17.40	20,880	10.44	1.74	1.49	"	2	1.50	1.63	1.16
39,200	19.60	23,520	11.76	1.96	1.58	"	2¼	1.68	1.73	1.23
43,600	21.80	26,160	13.08	2.18	1.66	"	2½	1.87	1.82	1.29
48,000	24.00	28,800	14.40	2.40	1.75	"	2¾	2.06	1.89	1.34
52,400	26.20	31,440	15.72	2.62	1.83	"	3	2.25	2.00	1.42
20,000	10.00	12,000	6.00	1.00	1.13	1	1	0.75	1.39	0.80
25,000	12.50	15,000	7.50	1.25	1.26	"	1¼	0.93	1.45	0.98
30,000	15.00	18,000	9.00	1.50	1.39	"	1½	1.12	1.52	1.08
35,000	17.50	21,000	10.50	1.75	1.49	"	1¾	1.31	1.64	1.16
40,000	20.00	24,000	12.00	2.00	1.60	"	2	1.50	1.75	1.24
45,000	22.50	27,000	13 50	2.25	1.70	"	2¼	1.68	1.86	1.32
50,000	25.00	30,000	15.00	2.50	1.79	"	2½	1.87	1.96	1.39
55,000	27.50	33,000	16.50	2.75	1.87	"	2¾	2.06	2.05	1.45
60,000	30.00	36,000	18.00	3.00	1.96	"	3	2.25	2.15	1.52
28,000	14.00	16,800	8.40	1.40	1.34	1⅛	1¼	0.93	1.47	1.04
33,600	16.80	20,160	10.08	1.68	1.47	"	1½	1.12	1.60	1.13
39,600	19.80	23,520	11.76	1.98	1.58	"	1¾	1.31	1.73	1.23
45,000	22.50	27,000	13.50	2.25	1.69	"	2	1.50	1.86	1.32
50,600	25.30	30,360	15.18	2.53	1.80	"	2¼	1.68	1.97	1.39
56,200	28.10	33,720	16.86	2.81	1.89	"	2½	1.87	2.09	1.48
61,800	30.90	37,080	18.54	3.09	1.98	"	2¾	2.06	2.18	1.54
67,400	33.70	40,440	20.22	3.37	2.08	"	3	2.25	2.26	1.60
73,000	36.50	43,800	21.90	3.65	2.16	"	3¼	2.43	2.36	1.67
78,600	39.30	47,160	23.58	3.93	2.24	"	3½	2.62	2.45	1.74
84,200	42.10	50,520	25.26	4.21	2.32	"	3¾	2 81	2.53	1.80
90,000	45.00	54,000	27.00	4.50	2.40	"	4	3.00	2.63	1.86
31,200	15.60	18,720	9.36	1.56	1.41	1¼	1¼	0.93	1.54	1.09
37,400	18.70	22,440	11.22	1.87	1.55	"	1½	1.12	1.69	1.20
43,600	21.80	26,160	13.08	2.18	1.67	"	1¾	1.31	1.82	1.29
50,000	25.00	30,000	15.00	2.50	1.79	"	2	1.50	1.96	1.39
56,200	28.10	33,720	16.86	2.81	1.89	"	2¼	1.68	2.09	1.48
62,400	31.20	37,440	18.72	3.12	1.99	"	2½	1.87	2.19	1.55

Capacity of tie or bar. 3 times safety. Lbs.	Capacity of tie or bar. 3 times safety. Tons.	Capacity of tie or bar. 5 times safety. Lbs.	Capacity of tie or bar. 5 times safety. Tons.	Sectional area in sq. inches.	Diameter in inches, if round.	Dimension of flat bars in in., uniform thickness. Thickness t of bar.	Dimension of flat bars in in., uniform thickness. Width b_1 of bar.	Dimension of flat bars in in., uniform thickness. Width b around eye.	Diameter D of pin or bolt. One place of shearing.	Diameter D of pin or bolt. Two places of shearing.
68,600	34.30	41,160	20.58	3.43	2.10	1¼	2¾	2.06	2.29	1.62
75,000	37.50	45,000	22.50	3.75	2.19	"	3	2.25	2.40	1.70
81,200	40.60	48,720	24.36	4.06	2.27	"	3¼	2.43	2.49	1.76
87,400	43.70	52,440	26.22	4.37	2.36	"	3½	2.62	2.60	1.84
93,600	46.80	56,160	28.08	4.68	2.44	"	3¾	2.81	2.68	1.89
100,000	50.00	60,000	30.00	5.00	2.53	"	4	3.00	2.77	1.96
41,200	20.60	24,720	12.36	2.06	1.62	1⅜	1½	1.12	1.77	1.26
48,000	24.00	28,800	14.40	2.40	1.75	"	1¾	1.31	1.89	1.34
55,000	27.50	33,000	16.50	2.75	1.87	"	2	1.50	2.05	1.45
61,800	30.90	37,080	18.54	3.09	1.98	"	2¼	1.68	2.18	1.54
68,600	34.30	41,160	20.58	3.43	2.09	"	2½	1.87	2.29	1.62
75,600	37.80	45,360	22.68	3.78	2.19	"	2¾	2.06	2.41	1.71
82,400	41.20	49,440	24.72	4.12	2.29	"	3	2.25	2.51	1.78
89,200	44.60	53,520	26.76	4.46	2.38	"	3¼	2.43	2.61	1.85
96,200	48.10	57,720	28.86	4.81	2.47	"	3½	2.62	2.71	1.92
103,000	51.50	61,800	30.90	5.15	2.56	"	3¾	2.81	2.81	1.99
110,000	55.00	66,000	33.00	5.50	2.65	"	4	3.00	2.90	2.05
45,000	22.5	27,000	13.50	2.25	1.70	1½	1½	1.12	1.86	1.32
52,400	26.20	31.440	15.72	2.62	1.83	"	1¾	1.31	2.00	1.42
60,000	30.00	36,000	18.00	3.00	1.96	"	2	1.50	2.15	1.52
67,400	33.70	40,440	20.22	3.37	2.07	"	2¼	1.68	2.27	1.61
75,000	37.50	45,000	22.50	3.75	2.19	"	2½	1.87	2.40	1.70
82,400	41.20	49,440	24.72	4.12	2.29	"	2¾	2.06	2.51	1.78
90,000	45.00	54,000	27.00	4.50	2.40	"	3	2.25	2.63	1.86
97,400	48.70	58.440	29.22	4.87	2.49	"	3¼	2.43	2.73	1.93
105,000	52.50	63,000	31.50	5.25	2.59	"	3½	2.62	2.84	2.01
113,400	56.20	67,440	33.72	5.62	2.67	"	3¾	2.81	2.93	2.08
120,000	60.00	72,000	36.00	6.00	2.77	"	4	3.00	3.03	2.15
127,400	63.70	76.440	38.22	6.37	2.85	"	4¼	3.18	3.12	2.21
135 000	67.50	81,000	40.50	6.75	2.93	"	4½	3.37	3.22	2.28
142,400	71.20	85,440	42.72	7.12	3.01	"	4¾	3.55	3.30	2.34
150,000	75 00	90,000	45.00	7.50	3.10	"	5	3.75	3.39	2.40

JOINTS OR CONNECTIONS IN IRON CONSTRUCTION.

Proportions of Bolts, Nuts, Rivets, &c.

Reference.

A = Sectional area of bolt, rivet, or pin.
A_1 = Sectional area of all rivets in a joint.
A_2 = Sectional area of one plate.
D = Diameter of bolt, rivet, or pin.
S = Ultimate resistance to shearing of material.
T = Ultimate resistance to tearing of material.
T_1 = Tensional strain on joint, &c.
a = Number of times that a bolt, &c., will have to be sheared. (See 2 on *Fig.* 258.)
d = Distance between centres of rivets.
k = Factor of safety.
l = Overlap, approximately $1\frac{2}{3}\ d$ to $1\frac{3}{4}\ d$.
m = Number of rivets in a joint.
n = Number of lines of rivets in a joint at right angles to strain.
t = Thickness of a plate.

Rivets.

Fig. 257.

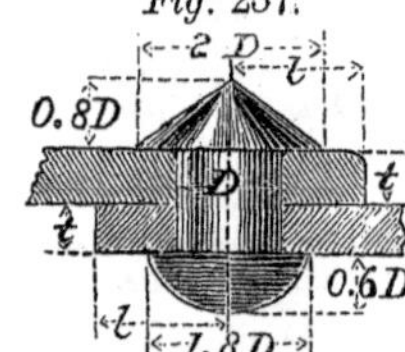

For tension in direction of rivet:

$$D = \sqrt{\frac{T_1\ k}{T\ 0.7854}}$$

For shearing at right angles:

If at one place $D = \sqrt{\frac{T_1\ k}{S\ 0.7854}}$

If at two places $D = \sqrt{\frac{T_1\ k}{S\ 1.5708}}$

Approximately: $l = 3t$ $D = 3t$

PIN, &c., IN TIE BARS.

Fig. 258.

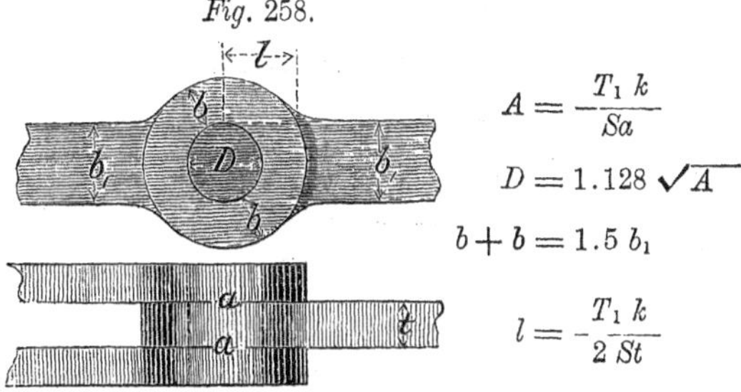

$$A = \frac{T_1 k}{Sa}$$

$$D = 1.128 \sqrt{A}$$

$$b + b = 1.5\, b_1$$

$$l = \frac{T_1 k}{2\, St}$$

PLATE JOINTS.

No. 1.—*Plate Joint Overlapped, four lines of Rivets.*

Fig. 259.

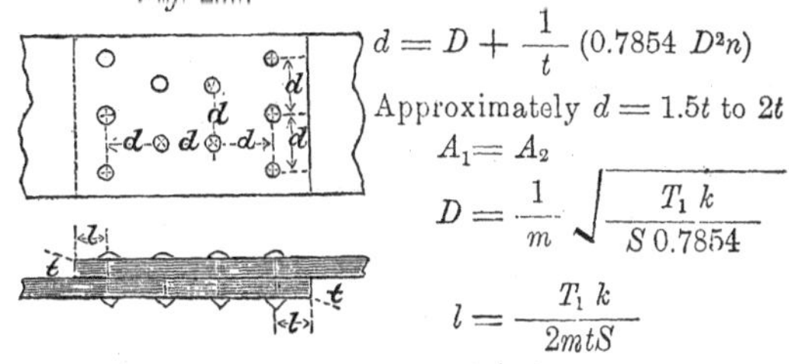

$$d = D + \frac{1}{t}\,(0.7854\, D^2 n)$$

Approximately $d = 1.5t$ to $2t$

$$A_1 = A_2$$

$$D = \frac{1}{m} \sqrt{\frac{T_1 k}{S\, 0.7854}}$$

$$l = \frac{T_1 k}{2mtS}$$

No. 2.—*Plate Joint Overlapped, single line of Rivet.*

Fig. 260. (Same as No. 1.)

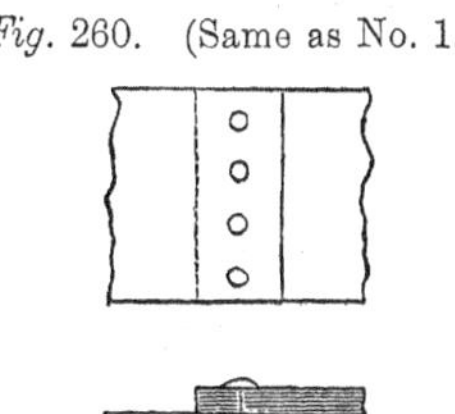

No. 3.—*Plate Joint Overlapped, two lines of Rivets.*

Fig. 261. (Same as No. 1.)

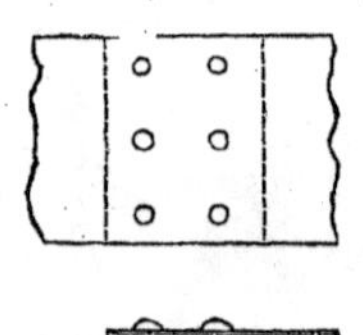

No. 4.—*Fish Joints, two lines of Rivets.*

Fig. 262.

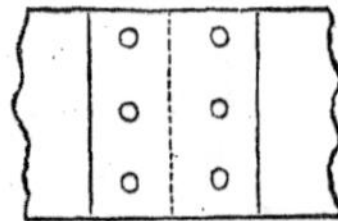

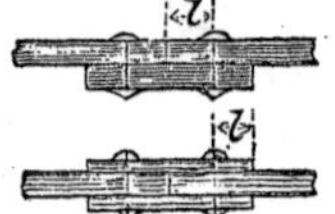

One fish plate. (Same as No. 1.)

Two fish plates.
Thickness of each fish plate $= \frac{1}{2}\, t$.

$$D = \frac{1}{m} \sqrt{\frac{T_1\, k}{S\, 1.5708}}$$

(Otherwise same as No. 1.)

DIMENSIONS OF BOLTS AND NUTS.

(Whitworth's proportions.)

Figs. 263, 264, 265, 266, 267, 268, 269, 270, and 271.

Dia. of Bolt.	Dimension of Nuts and Heads.					Dia. of Core.	No. Threads per inch.	
Inch.	Inch.	Inch.	Inch.	Inch.	Inch.	Inch.		
3	4½	5.18	5	7.07	3	2.57	3.5	1.50
2¾	4⅛	4.76	4½	6.37	2¾	2.35	3.5	1.75
2½	3¾	4.33	4⅛	5.83	2½	2.13	4.0	2.00
2¼	3⅜	3.89	3¾	5.30	2¼	1.91	4.0	2.12
2	3	3.46	3⅜	4.76	2	1.69	4.5	2.25
1⅞	2¾	3.17	3	4.24	1⅞	1.58	4.5	2.37
1¾	2⅝	3.03	2¾	3.89	1¾	1.47	5.0	2.50
1⅝	2½	2.88	2⅝	3.71	1⅝	1.36	5.0	2.75
1½	2¼	2.59	2½	3.53	1½	1.25	6.0	3.00
1⅜	2	2.30	2¼	3.18	1⅜	1.14	6.0	3.25
1¼	1⅞	2.16	2	2.82	1¼	1.08	7.0	3.50
1⅛	1⅝	1.87	1⅞	2.64	1⅛	0.92	7.0	4.00
1	1½	1.73	1⅝	2.29	1	0.81	8.0	5.00
⅞	1 5/16	1.51	1½	2.12	⅞	0.70	9.0	6.00
¾	1 3/16	1.38	1 5/16	1.86	¾	0.59	10.0	6.00
⅝	1	1.15	1 3/16	1.67	⅝	0.48	11.0	7.00
9/16	⅞	1.01	1	1.41	9/16	0.42	11.0	7.00
½	¾	0.86	⅞	1.23	½	0.37	12.0	8.00
7/16	¾	0.86	¾	1.06	7/16	0.31	14.0	8.00
⅜	9/16	0.64	¾	1.06	⅜	0.26	16.0	9.00
5/16	7/16	0.50	9/16	0.79	5/16	0.20	18.0	9.00
¼	⅜	0.43	9/16	0.79	¼	0.15	20.0	10.00

Fig. 272.

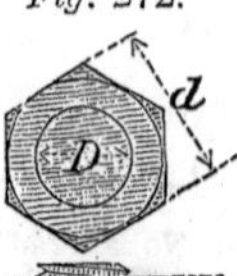

Approximate proportions of bolts, nuts, and heads in inches:

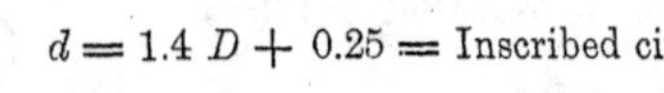

$d = 1.4\ D + 0.25$ = Inscribed circle.

$h = D$ = Height of nut.

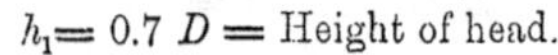

$h_1 = 0.7\ D$ = Height of head.

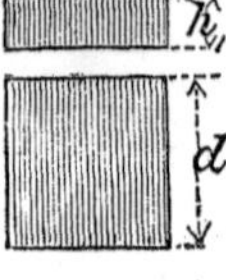

COMPOUND STRAINS IN HORIZONTAL AND SLOPING BEAMS.

(Load equally distributed or between supports.)

Area of Cross-section necessary to resist a Cross-breaking and Compressive Strain in Beams acting as a Boom in Trusses, &c., or Beams acting as Rafters, &c.

Reference.

m = Bending moment (See Page 100.)
C = Compressive strain. (See Roof and Simple Trusses.)
q = A factor depending on form of cross-section.
I = Moment of inertia of cross-section.
s = Distance from neutral axis to most compressed fibres.
A = Sectional area of beam, &c.
h = Depth of beam, &c.
p = Resistance to compression with safety per square inch of section.
W = Total load.
l = Length of beam, &c.

$$q = \frac{I}{\frac{s}{h}\, h^2\, A}$$

For horizontal beams, &c.:

$$A = \frac{1}{p}\left(\frac{M}{qh} + C\right) \qquad p = \frac{1}{A}\left(\frac{M}{qh} + C\right)$$

For sloping beams, &c., v = angle between horizontal and beam:

$$A = \frac{W}{p}\left[\frac{1}{2}\left(\frac{1}{\sin. v} + \sin. v\right) + \frac{l \cos. v}{12\, qh}\right]$$

$$p = \frac{W}{A}\left[\frac{1}{2}\left(\frac{1}{\sin. v} + \sin. v\right) + \frac{l \cos. v}{12\, qh}\right]$$

RAFTER OF A ROOF TRUSS.

Fig. 273.

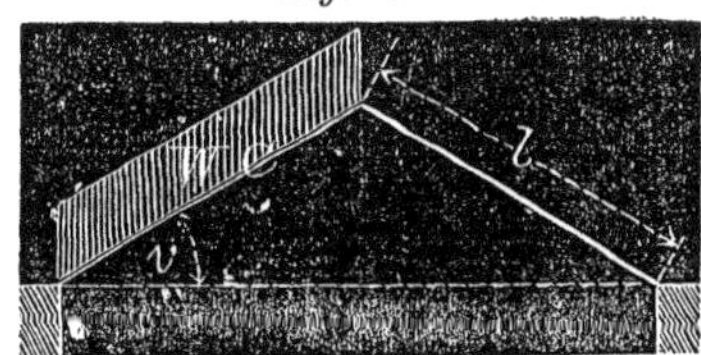

EXAMPLE.

Reference.

$W = 2.5$ tons. $C = 2.8$ tons. $l = 10$ feet. $v = 26° 30'$
$p = 5$ tons per square inch.

We will assume a Phœnix Co's six-inch beam of the following dimensions: $h = 6$ inches; $A = 4$ inches; $I = 22.5$

$$q = \frac{22.5}{0.5 \times 6^2 \times 4} = 0.312$$

$$A = \frac{2.5}{5}\left[\frac{1}{2}\left(\frac{1}{0.446} + 0.446\right) + \frac{120 \times 0.895}{12 \times 0.312 \times 6}\right] = 3.06 \text{ ins.};$$

showing that the six-inch beam has a greater sectional area than required.

If the load is concentrated at the apex of roof, the compressive strain $C = 2.8$ tons, and the area necessary to resist this strain would be (taking p at five tons per square inch) $\frac{2.8}{5} = 0.56$ sq. inches, provided this is able to resist buckling.

By comparing this with the above result, it will be seen how much greater the sectional area will have to be to resist a cross-breaking strain, caused by the load being distributed. These remarks also apply to simple trusses.

SIMPLE TRUSS, (BEAM CONTINUOUS OVER STRUT.)

Fig. 274.

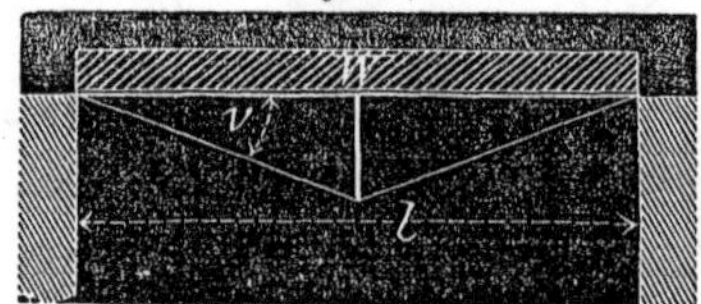

EXAMPLE.

Reference.

$W = 20$ tons. $l = 20$ feet. $v = 15°$ $p = 5$ tons per sq. inch.

We will assume a Phœnix Co's twelve-inch beam of the following dimensions:

$h = 12$ inches. $I = 275.92$
$A = 12.5$ inches. $s = 6$ inches.

$$q = \frac{275.92}{0.5 \times 12^2 \times 12.5} = 0.306$$

$m = 0.0703 \times 1 \times 120^2 = 84.36$ (See Reaction of Supports.)
$C = 23.32$ tons.

$$A = \frac{1}{5}\left(\frac{84.36}{0.306 \times 12}\right) + 23.32 = \frac{46.26}{5} = 9.25 \text{ inches.}$$

Consequently the sectional area of the twelve-inch beam is amply sufficient.

[NOTE.—The formulas for horizontal beams are also applicable to rafters of roof trusses, m and C being given. For the bending moments (m) the various distances are the horizontal projections of those on the rafter from abutment to ridge.

The foregoing formulas also apply to beams under a cross-breaking and tensional strain. If the truss (*Fig.* 274) is inverted, the horizontal member will be in tension. Hence, insert the resistance of the material to tension instead of compression, and put tensional for compressive strain; otherwise, the formulas remain the same.]

WEIGHT OF MOVING LOADS.

Variable and Accidental Loads.

(Weight of construction not included.)

Character of structure.	How loaded.	Weight in lbs. per square foot of surface.	
Street bridges for horse cars and heavy traffic.	Crow'd with persons.	Minimum.........	40 lbs.
		Maximum.........	120 "
		Average..........	80 "
Street bridges for general traffic, foot passengers, &c.	Persons, animals, and wagons.	Public travel....	80 lbs.
		Private travel...	40 "
		Heavy business wagons..........	80 "
		Light business wagons..........	40 "
Floors, &c........	Crowded public places.	Minimum.........	40 lbs.
		Maximum.........	120 "
		Average..........	80 "
	Dwellings..........	....	40 "
	Churches, court-rooms, theatres, and ball-rooms.	....	80 "
	Storage of grain...	....	100 "
	General merchandise..........	....	200 "
	Warehouses........	....	250 "
	Factories...........	....	200 to 400 "
	Hay-lofts...........	....	80 "

STATIC AND MOVING LOADS ON BRIDGES OF WROUGHT IRON.

The following table gives an approximate weight per lineal foot in pounds of the static load or weight of construction complete for *Single-Line Railway Bridges*, supported at the ends, from ten to four hundred feet span; also the weight of the moving load per lineal foot of span, based on the assumption that the heaviest locomotives exert a pressure of three thousand pounds per lineal foot between their extreme bearings.

The table is applicable in computing the strains in all trusses with parallel booms mentioned in this work.

Weight of Construction and Moving Load of Wrought-Iron Single-Line Railway Bridges for the heaviest traffic.

(From 20 to 400 feet span.)

Weight of construction complete, including cross-ties and rails.				Weight of moving load equal to 3,000 lbs. per lineal foot of load.			
Span in ft.	Weight in lbs. per lineal foot of span.	Span in ft.	Weight in lbs. per lineal foot of span.	Span in ft.	Weight in lbs. per lineal foot of span.	Span in ft.	Weight in lbs. per lin. foot of span.
10	427	210	1,891	10	6,300	210	2,535
20	500	220	1,964	20	5,370	220	2,495
30	573	230	2,037	30	4,250	230	2,455
40	646	240	2,110	40	3,780	240	2,375
50	719	250	2,183	50	3,550	250	2,335
60	792	260	2,256	60	3,400	260	2,290
70	865	270	2,329	70	3,300	270	2,245
80	938	280	2,402	80	3,250	280	2,200
90	1,011	290	2,475	90	3,180	290	2,160
100	1,084	300	2,548	100	3,120	300	2,120
110	1,157	310	2,621	110	3,050	310	2,080
120	1,230	320	2,694	120	3,000	320	2,045
130	1,303	330	2,767	130	2,930	330	2,010
140	1,380	340	2,840	140	2,880	340	1,975
150	1,453	350	2,913	150	2,820	350	1,940
160	1,526	360	2,986	160	2,760	360	1,910
170	1,599	370	3,059	170	2,700	370	1,880
180	1,672	380	3,132	180	2,655	380	1,850
190	1,745	390	3,205	190	2,615	390	1,820
200	1,818	400	3,278	200	2,575	400	1,800

The following gives the actual weight of some well-known Bridges (single line) in America, Germany, and England:

Name of Bridge.	System.	Span in feet.	Weight of construction per lineal foot. Lbs.	Weight of moving load per lineal foot. Lbs.	Strain in boom per square inch. Lbs.
"Brenz," near Königsbronn...	Open Web, parallel booms.	63.0	760	3,131	7,530
"Colomak"........	"	111.0	1,090	3,067	9,516
"Iser," near Munich	"	164.7	1,770	3,656	8,532
"Donau," near Ingolstadt......	"	178.0	1,954	3,312	8,532
"Elb," near Meissen...........	"	179.0	1,324	2,783	10,390
"Rhine," near Mainz............	"Pauli's," parabolic arched booms.	345.0	2,170	1,970	11,660
"Royal Albert," near Saltash...	"	455.0	4,418	2,240	9,954
"Boyne"	Lattice.............	264.0	3,225		
"Leven"...........	"	36.0	566		
"Kent"............	"	36.0	580		
"Harper's Ferry"	Truss.................	124.0	770		

MISCELLANEOUS.

GEOMETRY.

LONGIMETRY AND PLANIMETRY.

(Lines and Areas.)

Reference.

A = Area.
π = Periphery of circle = 3.14159 when diameter = 1.
r = Radius of circle.
c = Length of cord of segment.
p = Circumference of circle for given diameter.
l = Length of circle arc, &c.
h = Height of segment.
v = Angles, expressed in decimals, as $15^\circ\ 30' = 15.5$.
For other designations, see Figures.

[NOTE.—Always use the same unit for dimensions.]

Values of π.

$\pi = 1.14159$

$2\pi = 6.28319$

$\frac{1}{\pi} = 0.31831$

$\frac{1}{2\pi} = 0.15915$

$\frac{1}{\pi^2} = 0.10132$

$\frac{2}{\pi} = 0.63662$

$\frac{\pi}{2} = 1.57080$

$\frac{\pi}{3} = 1.04720$

$\frac{\pi}{4} = 0.78540$

$\frac{\pi}{6} = 0.52360$

$\pi^2 = 9.86960$

$\pi^3 = 31.00628$

$\sqrt{\pi} = 1.77245$

$\sqrt[3]{\pi} = 1.46459$

$\sqrt{\frac{1}{\pi}} = 0.56419$

Fig. 275.

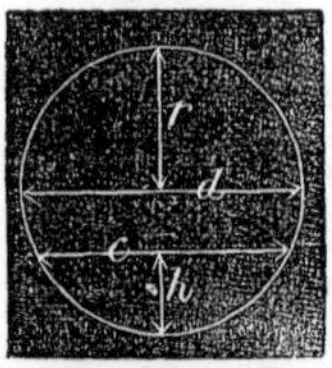

$$p = \pi d$$
$$d = \frac{p}{\pi}$$
$$r = \frac{p}{2\pi}$$

Fig. 276.

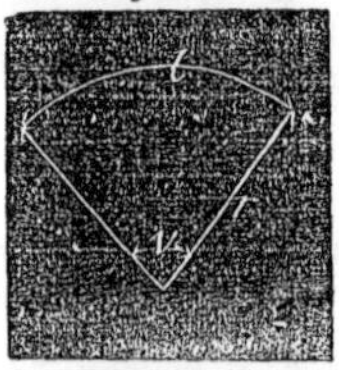

$$l = \frac{v}{360^\circ} p = \frac{v \pi d}{360^\circ} = \frac{v \pi r}{180^\circ}$$
$$v = \frac{l}{\pi r} 180^\circ$$
$$r = \frac{180^\circ}{v} \frac{l}{\pi}$$

Fig. 277.

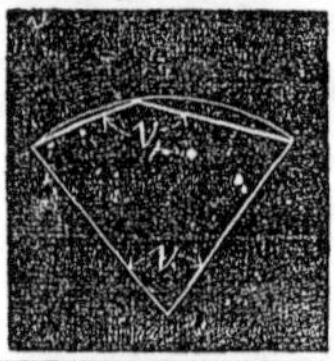

$$v_1 = 180^\circ - \frac{v}{2}$$
$$v = 2(180^\circ - v_1)$$

Fig. 278.

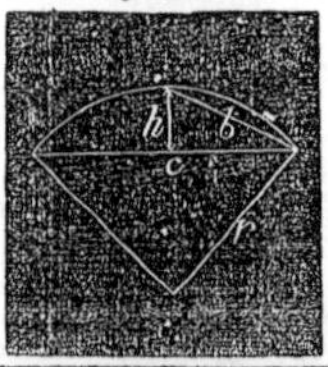

$$r = \frac{c^2 + 4h^2}{8h} = \frac{b^2}{2h}$$
$$c = 2\sqrt{2hr - h^2}$$
$$h = r - \sqrt{r^2 - \left(\frac{c}{2}\right)^2}$$

Fig. 279.

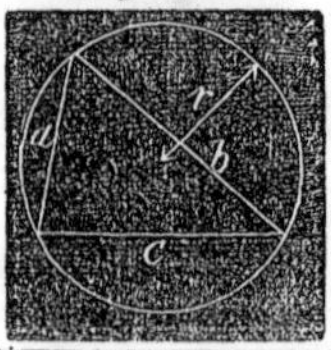

$$r = \frac{ac}{2\sqrt{a^2 - \left(\frac{a^2 + b^2 - c^2}{2b}\right)^2}}$$

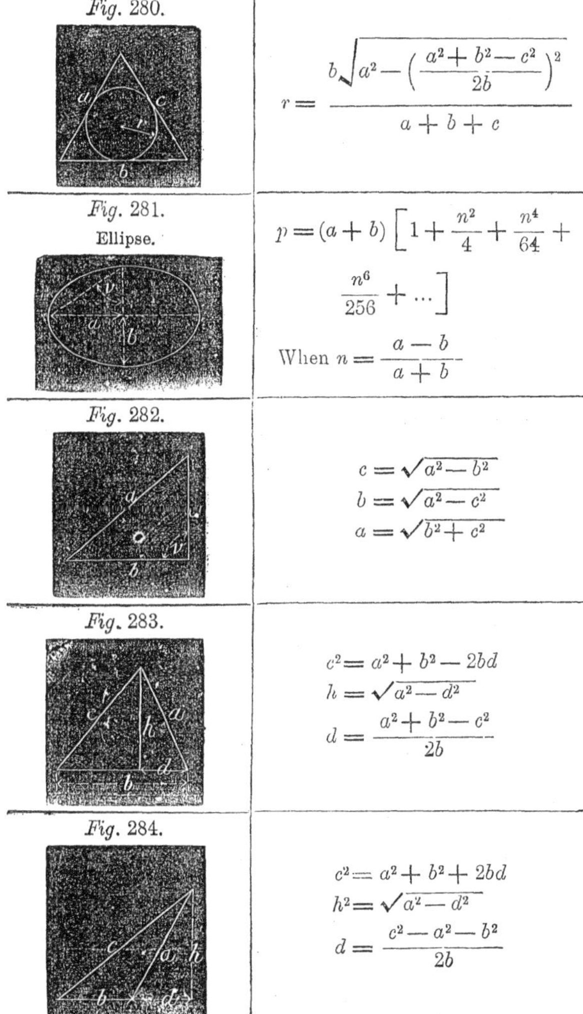

Fig. 280.	$r = \dfrac{b\sqrt{a^2 - \left(\dfrac{a^2 + b^2 - c^2}{2b}\right)^2}}{a + b + c}$
Fig. 281. Ellipse.	$p = (a + b)\left[1 + \dfrac{n^2}{4} + \dfrac{n^4}{64} + \dfrac{n^6}{256} + \ldots\right]$ When $n = \dfrac{a - b}{a + b}$
Fig. 282.	$c = \sqrt{a^2 - b^2}$ $b = \sqrt{a^2 - c^2}$ $a = \sqrt{b^2 + c^2}$
Fig. 283.	$c^2 = a^2 + b^2 - 2bd$ $h = \sqrt{a^2 - d^2}$ $d = \dfrac{a^2 + b^2 - c^2}{2b}$
Fig. 284.	$c^2 = a^2 + b^2 + 2bd$ $h^2 = \sqrt{a^2 - d^2}$ $d = \dfrac{c^2 - a^2 - b^2}{2b}$

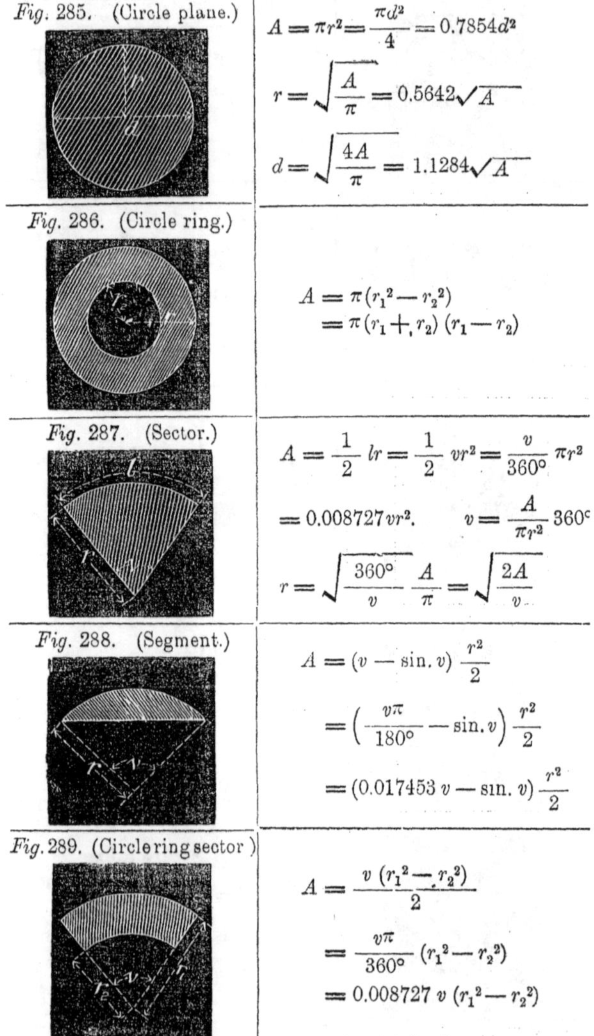

Fig. 285. (Circle plane.)

$$A = \pi r^2 = \frac{\pi d^2}{4} = 0.7854 d^2$$

$$r = \sqrt{\frac{A}{\pi}} = 0.5642 \sqrt{A}$$

$$d = \sqrt{\frac{4A}{\pi}} = 1.1284 \sqrt{A}$$

Fig. 286. (Circle ring.)

$$A = \pi (r_1^2 - r_2^2)$$
$$= \pi (r_1 + r_2)(r_1 - r_2)$$

Fig. 287. (Sector.)

$$A = \frac{1}{2} lr = \frac{1}{2} vr^2 = \frac{v}{360^\circ} \pi r^2$$

$$= 0.008727 vr^2. \qquad v = \frac{A}{\pi r^2} 360^\circ$$

$$r = \sqrt{\frac{360^\circ}{v} \frac{A}{\pi}} = \sqrt{\frac{2A}{v}}$$

Fig. 288. (Segment.)

$$A = (v - \sin. v) \frac{r^2}{2}$$

$$= \left(\frac{v\pi}{180^\circ} - \sin. v\right) \frac{r^2}{2}$$

$$= (0.017453\, v - \sin.\, v) \frac{r^2}{2}$$

Fig. 289. (Circle ring sector.)

$$A = \frac{v (r_1^2 - r_2^2)}{2}$$

$$= \frac{v\pi}{360^\circ} (r_1^2 - r_2^2)$$

$$= 0.008727\, v\, (r_1^2 - r_2^2)$$

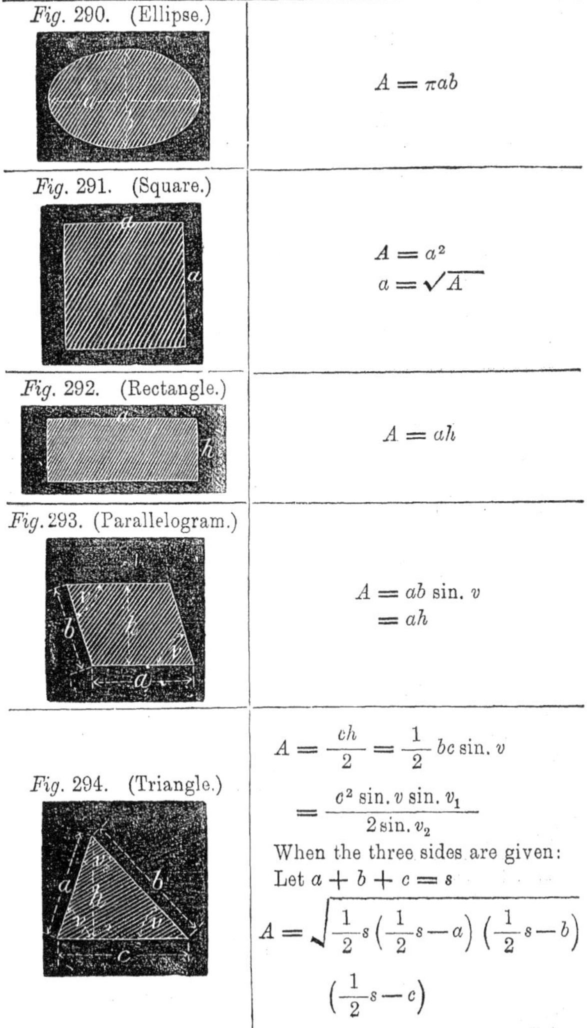

Fig. 290. (Ellipse.)	$A = \pi ab$
Fig. 291. (Square.)	$A = a^2$ $a = \sqrt{A}$
Fig. 292. (Rectangle.)	$A = ah$
Fig. 293. (Parallelogram.)	$A = ab \text{ sin. } v$ $= ah$
Fig. 294. (Triangle.)	$A = \frac{ch}{2} = \frac{1}{2} bc \text{ sin. } v$ $= \frac{c^2 \text{ sin. } v \text{ sin. } v_1}{2 \text{ sin. } v_2}$ When the three sides are given: Let $a + b + c = s$ $A = \sqrt{\frac{1}{2}s\left(\frac{1}{2}s - a\right)\left(\frac{1}{2}s - b\right)\left(\frac{1}{2}s - c\right)}$

CENTER OF GRAVITY OF PLANES.

Reference.

x = Distance from a fixed base to center of gravity.
r = Radius.
c = Chord.
b, p, h = Dimensions.
A = Area.
v = Angle.

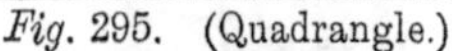 *Fig.* 295. (Quadrangle.) 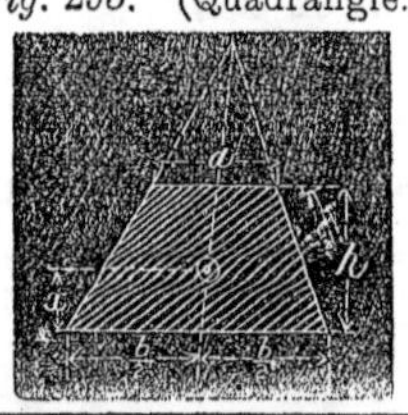	a and b parallel. $x = \frac{h}{2} - \frac{h}{6}\left(\frac{b-a}{b+a}\right)$
Fig. 296. (Triangle.) 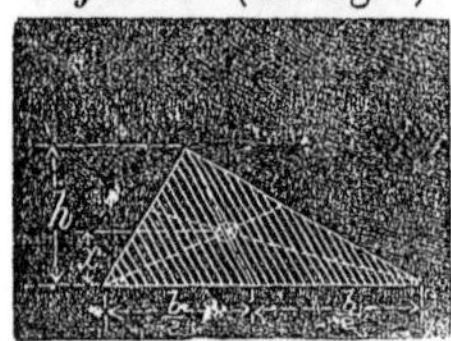	$x = \frac{h}{3}$
Fig. 297. (Half circle, or elliptic plane.) 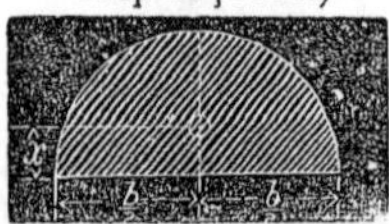	$\frac{b}{2}$ = radius $= r$ $x = 0.4244r$
Fig. 298. (Concentric ring.) 	$x = \frac{4}{3} \frac{\sin. \frac{1}{2}v}{v} \frac{r^3 - r_1^3}{r^2 - r_1^2}$

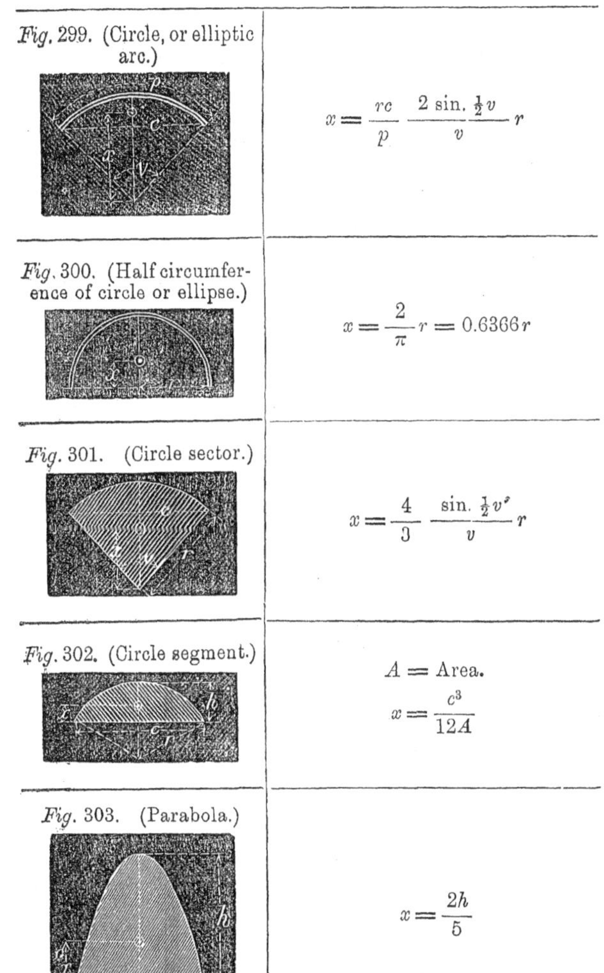

Fig. 299. (Circle, or elliptic arc.)	$x = \frac{rc}{p} \quad \frac{2 \sin. \frac{1}{2}v}{v} r$
Fig. 300. (Half circumference of circle or ellipse.)	$x = \frac{2}{\pi} r = 0.6366 r$
Fig. 301. (Circle sector.)	$x = \frac{4}{3} \quad \frac{\sin. \frac{1}{2}v}{v} r$
Fig. 302. (Circle segment.)	A = Area. $x = \frac{c^3}{12A}$
Fig. 303. (Parabola.)	$x = \frac{2h}{5}$

Fig. 305. (Half parabola.)

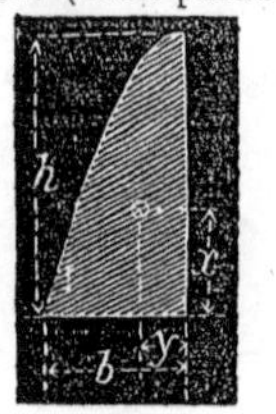

$$x = \frac{2}{5} h$$

$$y = \frac{3}{8} h$$

Of any section, composed of any number of simple figures:

Fig. 305.

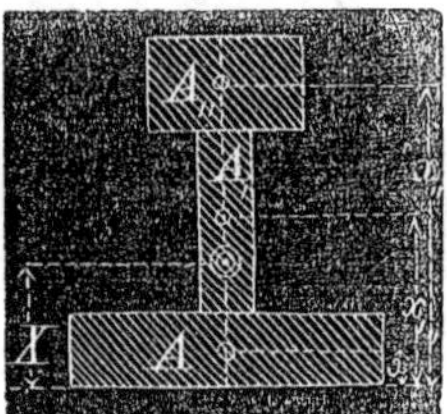

Additional Reference.

$A, A_{/}, A_{//}$ = Sectional area of simple figures.

X = Distance from center of gravity of whole section to axis *mn*.

$x, x_{/}, x_{//}$ = Distance from center of gravity of a simple figure to a fixed axis *mn*.

$$X = \frac{Ax + A_{/}x_{/} + A_{//}x_{/} + \&c.}{A\ A_{/} + A_{//} + \&c.}$$

TRIGONOMETRICAL FORMULAS.

Reference.

$a, b, c =$ Length of sides.
$A, B, C =$ Angles opposite to a, b, c respectively.

Right Angle Triangle.

Fig. 306.

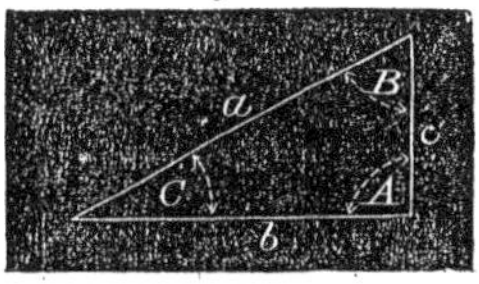

$$A = 90^\circ$$
$$a = \sqrt{b^2 + c^2}$$
$$a = \frac{c}{\text{sin. } C}$$
$$a = \frac{b}{\text{cos. } C}$$

$$b = a \text{ cos. } C$$
$$b = c \text{ cot. } C$$
$$b = a \text{ sin. } B$$
$$b = c \text{ tang. } B$$
$$c = b \text{ tang. } C$$
$$c = a \text{ sin. } C$$

$$\text{Sin. } C = \frac{c}{a}$$
$$\text{Cos. } C = \frac{b}{a}$$

$$\text{Tang. } C = \frac{c}{b} = \frac{\text{sin. } C}{\text{cos. } C} = \frac{1}{\text{cot. } C}$$
$$\text{Cotang. } C = \frac{\text{cos. } C}{\text{sin. } C} = \frac{1}{\text{tang. } C}$$
$$\text{Secant } C = \frac{1}{\text{cos. } C}$$
$$\text{Cosec. } C = \frac{1}{\text{sin. } C}$$

Oblique Angle Triangle.

Fig. 307.

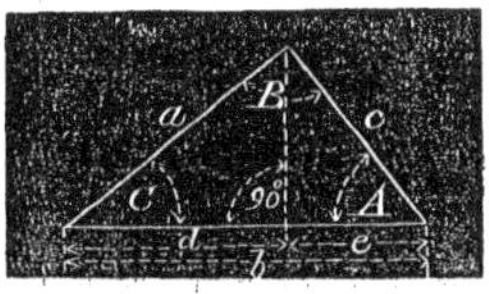

$$a = \frac{c \text{ sin. } A}{\text{sin. } C}$$
$$a = \frac{c \text{ sin. } A}{\text{sin. } (A + B)}$$

$$a = \sqrt{b^2 + c^2 - 2bc \text{ cos. } A}$$
$$b = \frac{c \text{ sin. } B}{\text{sin. } C}$$
$$d = \frac{1}{2}\left(b + \frac{(a+c)(a-c)}{b}\right)$$

$$e = \frac{1}{2}\left(b - \frac{(a+c)(a-c)}{b}\right)$$
$$\text{Sin. } C = \frac{c \text{ sin. } B}{b} = \frac{c \text{ sin. } A}{a}$$
$$\text{Sin. } A = \frac{a \text{ sin. } C}{c}$$

NATURAL SINE

Deg.	Minutes.						
	0	5	10	15	20	25	30
0	.00000	.00145	.00291	.00436	.00582	.00727	.00873
1	.01745	.01891	.02036	.02181	.02327	.02472	.02618
2	.03490	.03635	.03781	.03926	.04071	.04217	.04362
3	.05234	.05379	.05524	.05669	.05814	.05960	.06105
4	.06976	.07121	.07266	.07411	.07556	.07701	.07846
5	.08716	.08860	.09005	.09150	.09295	.09440	.09585
6	.10453	.10597	.10742	.10887	.11031	.11176	.11320
7	.12187	.12331	.12476	.12620	.12764	.12908	.13053
8	.13917	.14061	.14205	.14349	.14493	.14637	.14781
9	.15643	.15787	.15931	.16074	.16218	.16361	.16505
10	.17365	.17508	.17651	.17794	.17937	.18081	.18224
11	.19081	.19224	.19366	.19509	.19652	.19794	.19937
12	.20791	.20933	.21076	.21218	.21360	.21502	.21644
13	.22495	.22637	.22778	.22920	.23062	.23203	.23345
14	.24192	.24333	.24474	.24615	.24756	.24897	.25038
15	.25882	.26022	.26163	.26303	.26443	.26584	.26724
16	.27564	.27704	.27843	.27983	.28123	.28262	.28402
17	.29237	.29376	.29515	.29654	.29793	.29932	.30071
18	.30902	.31040	.31178	.31316	.31454	.31593	.31730
19	.32557	.32694	.32832	.32969	.33106	.33244	.33381
20	.34202	.34339	.34475	.34612	.34748	.34884	.35021
21	.35837	.35973	.36108	.36244	.36379	.36515	.36650
22	.37461	.37595	.37730	.37865	.37999	.38134	.38268
23	.39073	.39207	.39341	.39474	.39608	.39741	.39875
24	.40674	.40806	.40939	.41072	.41204	.41337	.41469
25	.42262	.42394	.42525	.42657	.42788	.42920	.43051
26	.43837	.43968	.44098	.44229	.44359	.44494	.44620
27	.45399	.45529	.45658	.45787	.45917	.46046	.46175
28	.46947	.47076	.47204	.47332	.47460	.47588	.47716
29	.48481	.48608	.48735	.48862	.48989	.49116	.49242
30	.50000	.50126	.50252	.50377	.50503	.50628	.50754
31	.51504	.51628	.51753	.51877	.52002	.52126	.52250
32	.52992	.53115	.53238	.53361	.53484	.53607	.53730
33	.54464	.54586	.54708	.54829	.54951	.55072	.55194
34	.55919	.56040	.56160	.56280	.56401	.56521	.56641
35	.57358	.57477	.57596	.57715	.57833	.57952	.58070
36	.58779	.58869	.59014	.59131	.59248	.59365	.59482
37	.60182	.60298	.60414	.60529	.60645	.60761	.60876
38	.61566	.61681	.61795	.61909	.62024	.62138	.62251
39	.62932	.63045	.63158	.63271	.63383	.63496	.63608
40	.64279	.64390	.64501	.64612	.64723	.64834	.64945
41	.65606	.65716	.65825	.65935	.66044	.66153	.66262
42	.66913	.67221	.67129	.67237	.67344	.67452	.67559
43	.68200	.68306	.68412	.68518	.68624	.68730	.68835
44	.69466	.69570	.69675	.69779	.69883	.69987	.70091
Deg.	60	55	50	45	40	35	30
	Minutes.						

NATURAL COSINE.

NATURAL SINE.

Minutes.						Deg.
35	40	45	50	55	60	
.01018	.01164	.01309	.01454	.01600	.01745	89
.02763	.02908	.03054	.03199	.03345	.03490	88
.04507	.04653	.04798	.04943	.05088	.05234	87
.06250	.06395	.06540	.06685	.06831	.06976	86
.07991	.08136	.08281	.08426	.08571	.08716	85
.09729	.09874	.10019	.10164	.10308	.10453	84
.11465	.11609	.11754	.11898	.12043	.12187	83
.13197	.13341	.13485	.13629	.13802	.13917	82
.14925	.15069	.15212	.15356	.15500	.15643	81
.16648	.16792	.16935	.17078	.17222	.17365	80
.18367	.18509	.18652	.18795	.18938	.19081	79
.20079	.20222	.20364	.20507	.20649	.20791	78
.21786	.21928	.22070	.22212	.22353	.22495	77
.23486	.23627	.23769	.23910	.24051	.24192	76
.25179	.25320	.25460	.25601	.25741	.25882	75
.26864	.27004	.27144	.27284	.27421	.27564	74
.28541	.28680	.28820	.28959	.29098	.29237	73
.30209	.30348	.30486	.30625	.30763	.30902	72
.31868	.32006	.32144	.32282	.32419	.32557	71
.33518	.33655	.33792	.33929	.34065	.34202	70
.35157	.35293	.35429	.35565	.35701	.35837	69
.36785	.36921	.37050	.37101	.37326	.37461	68
.38403	.38537	.38671	.38805	.38939	.39073	67
.40008	.40141	.40275	.40408	.40541	.40674	66
.41602	.41734	.41866	.41998	.42130	.42262	65
.43182	.43313	.43445	.43575	.43706	.43837	64
.44750	.44880	.45010	.45140	.45269	.45399	63
.46304	.46433	.46561	.46690	.46819	.46947	62
.47844	.47971	.48099	.48226	.48354	.48481	61
.49369	.49495	.49622	.49748	.49874	.50000	60
.50879	.51004	.51129	.51254	.51379	.51504	59
.52374	.52498	.52621	.52745	.52869	.52992	58
.53853	.53975	.54097	.54220	.54342	.54464	57
.55315	.55436	.55557	.55678	.55799	.55919	56
.56760	.56880	.57000	.57119	.57238	.57358	55
.58189	.58307	.58425	.58543	.58661	.58779	54
.59599	.59716	.59832	.59949	.60065	.60182	53
.60991	.61107	.61222	.61337	.61451	.61566	52
.62365	.62479	.62595	.62706	.62819	.62932	51
.63720	.63832	.63944	.64056	.64167	.64279	50
.65055	.65166	.65276	.65386	.65496	.65606	49
.66371	.66480	.66588	.66697	.66805	.66913	48
.67666	.67773	.67880	.67987	.68093	.68200	47
.68941	.69046	.69151	.69256	.69361	.69466	46
.70195	.70298	.70401	.70505	.70608	.70711	45
25	20	15	10	5	0	Deg.
Minutes.						

NATURAL COSINE.

NATURAL SINE.

Deg.	Minutes.						
	0	5	10	15	20	25	30
45	.70711	.70813	.70916	.71019	.71121	.71223	.71325
46	.71934	.72035	.72136	.72236	.72337	.72437	.72537
47	.73135	.73234	.73333	.73432	.73531	.73629	.73728
48	.74314	.74412	.74509	.74606	.74703	.74799	.74896
49	.75471	.75566	.75661	.75756	.75851	.75946	.76041
50	.76604	.76698	.76791	.76884	.76977	.77070	.77162
51	.77715	.77806	.77897	.77988	.78079	.78170	.78261
52	.78801	.78891	.78980	.79069	.79158	.79247	.79335
53	.79864	.79951	.80038	.80125	.80212	.80299	.80386
54	.80902	.80987	.81072	.81157	.81242	.81327	.81412
55	.81915	.81999	.82082	.82165	.82248	.82330	.82413
56	.82904	.82985	.83066	.83147	.83228	.83308	.83389
57	.83867	.83946	.84025	.84104	.84182	.84261	.84339
58	.84805	.84882	.84959	.85035	.85112	.85188	.85264
59	.85717	.85792	.85866	.85941	.86015	.86089	.86163
60	.86603	.86675	.86748	.86820	.86892	.86964	.87036
61	.87462	.87532	.87603	.87673	.87743	.87812	.87882
62	.88295	.88363	.88431	.88499	.88566	.88634	.88701
63	.89101	.89167	.89232	.89298	.89363	.89428	.89493
64	.89879	.89943	.90007	.90070	.90133	.90196	.90259
65	.90631	.90692	.90753	.90814	.90875	.90936	.90996
66	.91355	.91414	.91472	.91531	.91590	.91648	.91706
67	.92050	.92107	.92164	.92220	.92276	.92332	.92388
68	.92718	.92773	.92827	.92881	.92935	.92988	.93042
69	.93358	.93410	.93462	.93514	.93565	.93616	.93667
70	.93969	.94019	.94068	.94118	.94167	.94215	.94264
71	.94552	.94599	.94646	.94693	.94740	.94786	.94832
72	.95106	.95150	.95191	.95240	.95284	.95328	.95372
73	.95630	.95673	.95715	.95757	.95799	.95841	.95882
74	.96126	.96166	.96206	.96246	.96285	.96324	.96363
75	.96593	.96630	.96667	.96705	.96742	.96778	.96815
76	.97030	.97065	.97100	.97134	.97169	.97203	.97237
77	.97437	.97470	.97502	.97534	.97566	.97598	.97630
78	.97815	.97845	.97875	.97905	.97934	.97963	.97992
79	.98163	.98190	.98218	.98245	.98272	.98299	.98325
80	.98481	.98506	.98531	.98506	.98580	.98604	.98629
81	.98769	.98791	.98814	.98836	.98858	.98880	.98902
82	.99027	.99047	.99067	.99087	.99106	.99125	.99144
83	.99255	.99272	.99290	.99307	.99324	.99341	.99357
84	.99452	.99467	.99482	.99497	.99511	.99526	.99540
85	.99619	.99632	.99644	.99657	.99668	.99680	.99692
86	.99756	.99766	.99776	.99786	.99795	.99804	.99813
87	.99863	.99870	.99878	.99885	.99892	.99898	.99905
88	.99939	.99944	.99949	.99953	.99958	.99962	.99966
89	.99985	.99987	.99989	.99991	.99993	.99995	.99996
Deg.	60	55	50	45	40	35	30
	Minutes.						

NATURAL COSINE.

NATURAL SINE.

Minutes.						Deg.
35	40	45	50	55	60	
.71427	.71529	.71630	.71732	.71833	.71934	44
.72637	.72737	.72837	.72937	.73036	.73135	43
.73826	.73924	.74022	.74120	.74217	.74314	42
.74992	.75088	.75184	.75280	.75375	.75471	41
.76135	.76229	.76323	.76417	.76511	.76604	40
.77255	.77347	.77439	.77531	.77623	.77715	39
.78351	.78442	.78532	.78622	.78711	.78801	38
.79424	.79512	.79600	.79688	.79776	.79864	37
.80472	.80558	.80644	.80730	.80816	.80902	36
.81496	.81580	.81664	.81748	.81832	.81915	35
.82495	.82577	.82659	.82741	.82822	.82904	34
.83469	.83549	.83629	.83708	.83788	.83867	33
.84417	.84495	.84573	.84650	.84728	.84805	32
.85340	.85416	.85491	.85567	.85642	.85717	31
.86237	.86317	.86384	.86457	.86530	.86603	30
.87107	.87178	.87250	.87321	.87391	.87462	29
.87959	.88020	.88089	.88158	.88226	.88295	28
.88768	.88835	.88902	.88968	.89035	.89101	27
.89558	.89623	.89687	.89752	.89816	.89879	26
.90321	.90383	.90446	.90507	.90569	.90631	25
.91056	.91116	.91176	.91236	.91295	.91355	24
.91764	.91822	.91879	.91936	.91994	.92050	23
.92444	.92499	.92554	.92609	.92664	.92718	22
.93095	.93148	.93201	.93253	.93306	.93358	21
.93718	.93769	.93819	.93869	.93919	.93969	20
.94313	.94361	.94409	.94457	.94504	.94552	19
.94878	.94924	.94970	.95015	.95061	.95106	18
.95415	.95459	.95502	.95545	.95588	.95630	17
.95923	.95964	.96005	.96046	.96086	.96126	16
.96402	.96440	.96479	.96517	.96555	.96593	15
.96851	.96887	.96923	.96959	.96994	.97030	14
.97271	.97304	.97338	.97371	.97404	.97437	13
.97661	.97692	.97723	.97754	.97784	.97815	12
.98021	.98050	.98079	.98107	.98135	.98163	11
.98352	.98378	.98404	.98430	.98455	.98481	10
.98652	.98676	.98700	.98723	.98746	.98769	9
.98923	.98944	.98965	.98986	.99006	.99027	8
.99163	.99182	.99200	.99219	.99237	.99255	7
.99374	.99390	.99406	.99421	.99437	.99452	6
.99553	.99567	.99580	.99594	.99607	.99619	5
.99703	.99714	.99725	.99736	.99746	.99756	4
.99822	.99831	.99839	.99847	.99855	.99863	3
.99911	.99917	.99923	.99929	.99934	.99939	2
.99969	.99973	.99976	.99979	.99982	.99985	1
.99997	.99998	.99999	1.00000	1.00000	1.00000	0
25	20	15	10	5	0	Deg.
Minutes.						

NATURAL COSINE.

NATURAL TANGENT.

Deg.	Minutes.						
	0	5	10	15	20	25	30
0	0.0000	0.0014	0.0029	0.0044	0.0058	0.0073	0.0087
1	0.0175	0.0189	0.0204	0.0218	0.0233	0.0247	0.0262
2	0.0349	0.0364	0.0378	0.0393	0.0407	0.0422	0.0437
3	0.0524	0.0539	0.0553	0.0568	0.0582	0.0597	0.0612
4	0.0699	0.0714	0.0728	0.0743	0.0758	0.0772	0.0787
5	0.0875	0.0889	0.0904	0.0919	0.0933	0.0948	0.0963
6	0.1051	0.1066	0.1080	0.1095	0.1110	0.1125	0.1139
7	0.1228	0.1243	0.1257	0.1272	0.1287	0.1302	0.1316
8	0.1405	0.1420	0.1435	0.1450	0.1465	0.1480	0.1495
9	0.1584	0.1599	0.1614	0.1629	0.1644	0.1658	0.1673
10	0.1763	0.1778	0.1793	0.1808	0.1823	0.1838	0.1853
11	0.1944	0.1959	0.1974	0.1989	0.2004	0.2019	0.2034
12	0.2126	0.2141	0.2156	0.2171	0.2186	0.2202	0.2217
13	0.2309	0.2324	0.2339	0.2355	0.2370	0.2385	0.2401
14	0.2493	0.2509	0.2524	0.2540	0.2555	0.2571	0.2586
15	0.2679	0.2695	0.2711	0.2726	0.2742	0.2758	0.2773
16	0.2867	0.2883	0.2899	0.2915	0.2930	0.2946	0.2962
17	0.3057	0.3073	0.3089	0.3105	0.3121	0.3137	0.3153
18	0.3249	0.3265	0.3281	0.3297	0.3314	0.3330	0.3346
19	0.3443	0.3460	0.3476	0.3492	0.3508	0.3525	0.3541
20	0.3640	0.3656	0.3673	0.3689	0.3706	0.3722	0.3739
21	0.3839	0.3855	0.3872	0.3889	0.3905	0.3922	0.3939
22	0.4040	0.4057	0.4074	0.4091	0.4108	0.4125	0.4142
23	0.4245	0.4262	0.4279	0.4296	0.4314	0.4331	0.4348
24	0.4452	0.4470	0.4487	0.4505	0.4522	0.4540	0.4557
25	0.4663	0.4681	0.4698	0.4716	0.4734	0.4752	0.4770
26	0.4877	0.4895	0.4913	0.4931	0.4950	0.4968	0.4986
27	0.5095	0.5114	0.5132	0.5150	0.5169	0.5187	0.5206
28	0.5317	0.5336	0.5354	0.5373	0.5392	0.5411	0.5430
29	0.5543	0.5562	0.5581	0.5600	0.5619	0.5638	0.5658
30	0.5774	0.5793	0.5812	0.5832	0.5851	0.5871	0.5891
31	0.6008	0.6028	0.6048	0.6068	0.6088	0.6108	0.6128
32	0.6249	0.6269	0.6289	0.6309	0.6330	0.6350	0.6371
33	0.6494	0.6515	0.6535	0.6556	0.6577	0.6598	0.6619
34	0.6745	0.6766	0.6787	0.6809	0.6830	0.6851	0.6873
35	0.7002	0.7024	0.7045	0.7067	0.7089	0.7111	0.7133
36	0.7265	0.7288	0.7310	0.7332	0.7355	0.7377	0.7400
37	0.7536	0.7558	0.7581	0.7604	0.7627	0.7650	0.7673
38	0.7813	0.7836	0.7860	0.7883	0.7907	0.7931	0.7954
39	0.8098	0.8122	0.8146	0.8170	0.8195	0.8219	0.8243
40	0.8391	0.8416	0.8441	0.8466	0.8491	0.8516	0.8541
41	0.8693	0.8718	0.8744	0.8770	0.8795	0.8821	0.8847
42	0.9004	0.9030	0.9057	0.9083	0.9110	0.9137	0.9163
43	0.9325	0.9352	0.9380	0.9407	0.9434	0.9462	0.9490
44	0.9657	0.9685	0.9713	0.9742	0.9770	0.9798	0.9827
Deg.	60	55	50	45	40	35	30
	Minutes.						

NATURAL COTANGENT.

NATURAL TANGENT.

Minutes.						Deg.
35	40	45	50	55	60	
0.0102	0.0116	0.0131	0.0145	0.0160	0.0175	89
0.0276	0.0291	0.0305	0.0320	0.0335	0.0349	88
0.0451	0.0466	0.0480	0.0495	0.0509	0.0524	87
0.0626	0.0641	0.0655	0.0670	0.0685	0.0699	86
0.0802	0.0816	0.0831	0.0846	0.0860	0.0875	85
0.0978	0.0992	0.1007	0.1022	0.1036	0.1051	84
0.1154	0.1169	0.1184	0.1198	0.1213	0.1228	83
0.1331	0.1346	0.1361	0.1376	0.1391	0.1405	82
0.1509	0.1524	0.1539	0.1554	0.1569	0.1584	81
0.1688	0.1703	0.1718	0.1733	0.1748	0.1763	80
0.1868	0.1883	0.1899	0.1914	0.1929	0.1944	79
0.2050	0.2065	0.2080	0.2095	0.2110	0.2126	78
0.2232	0 2247	0.2263	0.2278	0.2293	0.2309	77
0.2416	0.2432	0.2447	0.2462	0.2478	0.2493	76
0.2602	0.2617	0.2633	0.2648	0.2664	0.2679	75
0.2789	0.2805	0.2820	0.2836	0.2852	0.2867	74
0.2978	0.2994	0.3010	0.3026	0.3041	0.3057	73
0.3169	0.3185	0.3201	0.3217	0.3233	0.3249	72
0.3362	0.3378	0.3394	0.3411	0.3427	0.3443	71
0.3558	0.3574	0.3590	0.3607	0.3623	0.3640	70
0.3755	0.3772	0.3789	0.3805	0.3822	0.3839	69
0.3956	0.3973	0.3990	0.4006	0.4023	0.4040	68
0.4159	0.4176	0.4193	0.4210	0.4228	0.4245	67
0.4365	0.4383	0.4400	0.4417	0.4435	0.4452	66
0.4575	0.4592	0.4610	0.4628	0.4645	0.4663	65
0.4788	0.4805	0.4823	0.4841	0.4859	0.4877	64
0.5004	0.5022	0.5040	0.5059	0.5077	0.5095	63
0.5224	0.5243	0.5261	0.5280	0.5298	0.5317	62
0.5448	0.5467	0.5486	0.5505	0.5524	0.5543	61
0.5677	0.5696	0.5715	0.5735	0.5754	0.5774	60
0.5910	0.5930	0.5949	0.5969	0.5989	0.6008	59
0.6148	0.6168	0.6188	0.6208	0.6228	0.6249	58
0.6391	0.6412	0.6432	0.6453	0.6473	0.6494	57
0.6640	0.6661	0.6682	0.6703	0.6724	0.6745	56
0.6894	0.6916	0.6937	0.6959	0.6980	0.7002	55
0.7155	0.7177	0.7199	0.7221	0.7243	0.7265	54
0.7422	0.7445	0.7467	0.7490	0.7513	0.7536	53
0.7696	0.7720	0.7743	0.7766	0.7789	0.7813	52
0.7978	0.8002	0.8026	0.8050	0.8074	0.8098	51
0.8268	0.8292	0.8317	0.8341	0.8366	0.8391	50
0.8566	0.8591	0.8617	0.8642	0.8667	0.8693	49
0.8873	0.8899	0.8925	0.8951	0.8978	0.9004	48
0.9190	0.9217	0.9244	0.9271	0.9298	0.9325	47
0.9517	0.9545	0.9573	0.9601	0 9629	0.9657	46
0.9856	0.9884	0.9913	0.9942	0.9971	1.0000	45
25	20	15	10	5	0	Deg.
Minutes.						

NATURAL COTANGENT.

NATURAL TANGENT.

Deg.	Minutes.						
	0	5	10	15	20	25	30
45	1.0000	1.0029	1.0058	1.0088	1.0117	1.0146	1.0176
46	1.0355	1.0385	1.0416	1.0446	1.0477	1.0507	1.0538
47	1.0724	1.0755	1.0786	1 0818	1.0850	1.0881	1.0913
48	1 1106	1.1139	1.1171	1.1204	1.1237	1.1270	1.1303
49	1.1504	1.1537	1.1571	1.1606	1.1640	1.1674	1.1708
50	1.1917	1.1953	1.1988	1.2024	1.2059	1.2095	1.2131
51	1.2349	1.2386	1.2423	1.2460	1.2497	1.2534	1.2572
52	1.2799	1.2838	1.2876	1.2915	1.2954	1.2993	1.3032
53	1.3270	1.3311	1.3351	1.3392	1.3432	1.3472	1.3514
54	1.3764	1.3806	1.3848	1.3891	1.3934	1.3976	1.4019
55	1.4281	1.4326	1.4370	1.4415	1.4460	1.4505	1.4550
56	1.4826	1.4872	1.4919	1.4966	1.5013	1.5061	1.5108
57	1.5399	1.5448	1.5497	1.5547	1.5597	1.5647	1.5697
58	1.6003	1.6055	1.6107	1.6160	1.6212	1.6265	1.6318
59	1.6643	1.6698	1.6753	1.6808	1.6864	1.6920	1.6976
60	1.7320	1.7379	1.7437	1.7496	1.7556	1.7615	1.7675
61	1.8040	1.8102	1.8165	1.8228	1.8291	1.8354	1.8418
62	1.8807	1.8873	1.8940	1.9007	1.9074	1.9142	1.9210
63	1.9626	1.9697	1.9768	1.9840	1.9912	1.9984	2.0057
64	2.0503	2.0579	2.0655	2.0732	2.0809	2.0887	2.0965
65	2.1445	2.1527	2.1609	2.1692	2.1775	2.1859	2.1943
66	2.2460	2.2549	2.2637	2.2727	2.2817	2.2907	2.2998
67	2.3558	2.3654	2.3750	2.3847	2.3945	2.4043	2.4142
68	2.4751	2.4855	2.4960	2.5065	2.5171	2.5279	2.5386
69	2.6051	2.6165	2.6279	2.6394	2.6511	2.6628	2.6746
70	2.7475	2.7600	2.7725	2.7852	2.7980	2.8109	2.8239
71	2.9042	2.9180	2.9319	2.9456	2.9600	2.9743	2.9886
72	3.0777	3.0930	3.1084	3.1240	3.1397	3.1556	3.1716
73	3.2708	3.2879	3.3052	3.3226	3.3402	3.3580	3.3759
74	3.4874	3.5067	3.5261	3.5457	3.5656	3.5856	3.6059
75	3.7320	3.7539	3.7760	3.7983	3.8208	3.8436	3.8667
76	4.0108	4.0358	4.0611	4.0867	4.1126	4.1388	4.1653
77	4.3315	4.3604	4.3897	4.4194	4.4494	4.4799	4.5107
78	4.7046	4.7385	4.7729	4.8077	4.8430	4.8788	4.9152
79	5.1445	5.1848	5.2257	5.2671	5.3093	5.3521	5.3955
80	5.6713	5.7199	5.7694	5.8197	5.8708	5.9228	5.9758
81	6.3137	6.3737	6.4348	6.4971	6.5605	6.6252	6.6912
82	7.1154	7.1912	7.2687	7.3479	7.4287	7.5113	7.5957
83	8.1443	8.2434	8.3450	8.4490	8.5555	8.6648	8.7769
84	9.5144	9.6493	9.7882	9.9310	10.0780	10.2290	10.3850
85	11.4300	11.6250	11.8260	12.0350	12.2510	12.4740	12.7060
86	14.3010	14.6060	14.9240	15.2570	15.6050	15.9690	16.3500
87	19.0810	19.6270	20.2060	20.8190	21.4700	22.1640	22.9040
88	28.6360	29.8820	31.2420	32.7300	34.3680	36.1780	38.1880
89	57.2900	62.4990	68.7500	76.3900	85.9480	98.2180	114.5900
Deg.	60	55	50	45	40	35	30
	Minutes.						

NATURAL COTANGENT.

NATURAL TANGENT.

Minutes.						Deg.
35	40	45	50	55	60	
1.0206	1.0235	1.0265	1.0295	1.0325	1.0355	44
1.0568	1.0590	1.0630	1.0661	1.0692	1.0724	43
1.0945	1.0977	1.1009	1.1041	1.1074	1.1106	42
1.1336	1.1369	1.1403	1.1436	1.1470	1.1504	41
1.1743	1.1778	1.1812	1.1847	1.1882	1.1917	40
1.2167	1.2203	1.2239	1.2276	1.2312	1.2349	39
1.2609	1.2647	1.2685	1.2723	1.2761	1.2799	38
1.3071	1.3111	1.3151	1.3190	1.3230	1.3270	37
1.3555	1.3597	1.3638	1.3680	1.3722	1·3764	36
1.4063	1.4106	1.4150	1.4193	1.4237	1.4281	35
1.4595	1.4641	1.4687	1.4733	1.4779	1.4826	34
1.5156	1.5204	1.5252	1.5301	1.5350	1.5399	33
1.5747	1.5798	1.5849	1.5900	1.5952	1.6003	32
1.6372	1.6426	1.6479	1.6534	1.6588	1.6643	31
1.7033	1.7090	1.7147	1.7205	1.7263	1.7320	30
1.7735	1.7795	1.7856	1.7917	1.7979	1.8040	29
1.8482	1.8546	1.8611	1.8676	1.8741	1.8807	28
1.9278	1.9347	1.9416	1.9486	1.9556	1.9626	27
2.0130	2.0204	2.0278	2.0353	2.0428	2.0503	26
2.1044	2.1123	2.1203	2.1283	2.1364	2.1445	25
2.2028	2.2113	2.2199	2.2286	2.2373	2.2460	24
2.3090	2.3183	2.3276	2.3369	2.3464	2.3558	23
2.4242	2.4342	2.4410	2.4545	2.4648	2.4751	22
2.5495	2.5605	2.5715	2.5826	2.5938	2.6051	21
2.6865	2.6985	2.7106	2.7228	2.7351	2.7475	20
2.8370	2.8502	2.8636	2.8770	2.8905	2.9042	19
3.0032	3.0178	3.0326	3.0475	3.0625	3.0777	18
3.1877	3.2041	3.2205	3.2371	3.2539	3.2708	17
3.3941	3.4124	3.4308	3.4495	3.4684	3.4874	16
3.6264	3.6471	3.6680	3.6891	3.7105	3.7320	15
3.8900	3.9136	3.9375	3.9616	3.9861	4.0108	14
4.1921	4.2193	4.2468	4.2747	4.3029	4.3315	13
4.5420	4.5736	4.6057	4.6382	4.6712	4.7046	12
4.9520	4.9894	5.0273	5.0658	5.1049	5.1445	11
5.4397	5.4845	5.5301	5.5764	5.6234	5.6713	10
6.0296	6.0844	6.1402	6.1970	6.2549	6.3137	9
6.7584	6.8269	6.8969	6.9682	7.0410	7.1154	8
7.6821	7.7703	7.8606	7.9530	8.0476	8.1443	7
8.8918	9.0098	9.1309	9.2553	9.3831	9.5144	6
10.5460	10.7120	10.8830	11.0590	11.2420	11.4300	5
12.9470	13.1970	13.4570	13.7270	14.0080	14.3010	4
16.7500	17.1690	17.6110	18.0750	18.5640	19.0810	3
23.6940	24.5420	25.4520	26.4320	27.4900	28.6360	2
40.4360	42.9640	45.8290	49.1040	52.8820	57.2900	1
137.5100	171.8800	229.1800	343.7700	687.5500		0
25	20	15	10	5	0	Deg.
Minutes.						

NATURAL COTANGENT.

NATURAL SECANT.

Deg.	Minutes.						
	0	5	10	15	20	25	30
0	1.0000	1.0000	1.0000	1.0000	1.0000	1.0000	1.0000
1	1.0001	1.0002	1.0002	1.0002	1.0003	1.0003	1.0003
2	1.0006	1.0007	1.0007	1.0008	1.0008	1.0009	1.0009
3	1.0014	1.0014	1.0015	1.0016	1.0017	1.0018	1.0019
4	1.0024	1.0025	1.0026	1.0027	1.0029	1.0030	1.0031
5	1.0038	1.0039	1.0041	1.0042	1.0043	1.0045	1.0046
6	1.0055	1.0057	1.0058	1.0060	1.0061	1.0063	1.0065
7	1.0075	1.0077	1.0079	1.0080	1.0082	1.0084	1.0086
8	1.0098	1.0100	1.0102	1.0104	1.0107	1.0109	1.0111
9	1.0125	1.0127	1.0129	1.0132	1.0134	1.0136	1.0139
10	1.0154	1.0157	1 0159	1.0162	1.0165	1.0167	1.0170
11	1.0187	1.0190	1.0193	1.0196	1.0199	1.0202	1.0205
12	1 0223	1.0226	1.0229	1.0233	1.0236	1.0239	1.0243
13	1.0263	1.0266	1.0270	1.0274	1.0277	1.0280	1.0284
14	1.0306	1.0310	1.0314	1.0317	1.0321	1.0325	1.0329
15	1.0353	1.0357	1.0361	1.0365	1.0369	1.0373	1.0377
16	1.0403	1.0407	1.0412	1.0416	1.0420	1.0425	1.0429
17	1.0457	1.0461	1.0466	1.0471	1.0476	1.0480	1.0485
18	1.0515	1.0520	1.0525	1.0530	1.0535	1.0540	1.0545
19	1.0577	1.0581	1.0587	1.0592	1.0598	1.0603	1.0608
20	1.0642	1.0647	1.0653	1.0659	1.0664	1.0670	1.0676
21	1.0711	1.0717	1.0723	1.0729	1.0736	1.0742	1.0748
22	1.0785	1.0792	1.0798	1.0804	1.0811	1.0817	1.0824
23	1.0864	1.0870	1.0877	1.0884	1.0891	1.0897	1.0904
24	1.0946	1.0953	1.0961	1.0968	1.0975	1.0982	1.0989
25	1.1034	1.1041	1.1049	1.1056	1.1064	1.1072	1.1079
26	1.1126	1.1134	1.1142	1.1150	1.1158	1.1166	1.1174
27	1.1223	1.1231	1.1240	1.1248	1.1257	1.1265	1.1274
28	1.1326	1.1334	1.1343	1.1352	1.1361	1.1370	1.1379
29	1.1433	1.1443	1.1452	1.1461	1.1471	1.1480	1.1489
30	1.1547	1.1557	1·1566	1.1576	1.1586	1.1596	1.1606
31	1.1666	1.1676	1.1687	1.1697	1.1707	1.1718	1.1728
32	1.1792	1.1802	1.1830	1.1824	1.1835	1.1846	1.1857
33	1.1923	1.1935	1.1946	1.1958	1.1969	1.1980	1.1992
34	1.2062	1.2074	1.2068	1.2098	1.2110	1.2122	1.2134
35	1.2208	1.2220	1.2233	1.2245	1.2258	1.2270	1.2283
36	1.2361	1.2374	1.2387	1.2400	1.2413	1.2427	1.2440
37	1.2521	1.2535	1.2549	1.2563	1.2577	1.2591	1.2605
38	1.2690	1.2705	1.2719	1.2734	1.2748	1.2763	1.2778
39	1.2867	1.2883	1.2898	1 2913	1.2929	1.2944	1.2960
40	1.3054	1.3070	4.3086	1.3102	1.3118	1.3134	1.3151
41	1.3250	1.3267	1.3284	1.3301	1.3318	1.3335	1.3352
42	1.3456	1.3474	1.3492	1.3509	1.3507	1.3540	1.3563
43	1.3673	1.3692	1.3710	1.3729	1.3748	1.3767	1.3786
44	1.3902	1.3921	1.3941	1.3960	1.3980	1.4000	1.4020
Deg.	60	55	50	45	40	35	30
	Minutes.						

NATURAL COSECANT.

NATURAL SECANT.

Minutes.						Deg.
35	40	45	50	55	60	
1.0000	1.0001	1.0001	1.0001	1.0001	1.0001	89
1.0004	1.0004	1.0005	1.0005	1.0005	1.0006	88
1.0010	1.0011	1.0011	1.0012	1.0013	1.0014	87
1.0019	1.0020	1.0021	1.0022	1.0023	1.0024	86
1.0032	1.0033	1.0034	1.0036	1.0037	1.0038	85
1.0048	1.0049	1.0050	1.0052	1.0053	1.0055	84
1.0066	1.0068	1.0070	1.0071	1.0073	1.0075	83
1.0088	1.0090	1.0092	1.0094	1.0096	1.0098	82
1.0113	1.0115	1.0118	1.0120	1.0122	1.0125	81
1.0141	1.0145	1.0146	1.0149	1.0152	1.0154	80
1.0173	1.0176	1.0179	1.0181	1.0184	1.0187	79
1.0208	1.0211	1.0214	1.0217	1.0220	1.0223	78
1.0246	1.0249	1.0253	1.0256	1.0260	1.0263	77
1.0288	1.0291	1.0295	1.0298	1.0302	1.0306	76
1.0333	1.0337	1.0341	1.0345	1.0349	1.0353	75
1.0382	1.0386	1.0390	1.0394	1.0399	1.0403	74
1.0434	1.0438	1.0443	1.0448	1.0452	1.0457	73
1.0490	1.0495	1.0500	1.0505	1.0510	1.0515	72
1.0550	1.0555	1.0560	1.0565	1.0571	1.0577	71
1.0644	1.0619	1.0625	1.0630	1.0636	1.0642	70
1.0682	1.0688	1.0694	1.0699	1.0705	1.0711	69
1.0754	1.0760	1.0766	1.0773	1.0779	1.0785	68
1.0830	1.0837	1.0844	1.0850	1.0857	1.0864	67
1.0911	1.0918	1.0925	1.0932	1.0939	1.0946	66
1.0997	1.1004	1.1011	1.1019	1.1026	1.1034	65
1.1087	1.1095	1.1102	1.1110	1.1118	1.1126	64
1.1182	1.1190	1.1198	1.1207	1.1215	1.1223	63
1.1282	1.1291	1.1299	1.1308	2.1317	1.1326	62
1.1388	1.1397	1.1406	1.1415	1.1424	1.1433	61
1.1499	1.1508	1.1518	1.1528	1.1537	1.1547	60
1.1616	1.1626	1.1636	1.1646	1.1656	1.1666	59
1.1739	1.1749	1.1760	1.1770	1.1781	1.1792	58
1.1868	1.1879	1.1819	1.1901	1.1912	1.1923	57
1.2004	1.2015	1.2027	1.2039	1.2050	1.2062	56
1.2146	1.2158	1.2171	1.2183	1.2195	1.2208	55
1.2296	1.2309	1.2322	1.2335	1.2348	1.2361	54
1.2453	1.2467	1.2480	1.2494	1.2508	1.2521	53
1.2619	1.2633	1.2647	1.2661	1.2676	1.2690	52
1.2793	1.2807	1.2822	1.2837	1.2852	1.2867	51
1.2975	1.2991	1.3006	1.3022	1.3038	1.3054	50
1.3167	1.3184	1.3200	1.3217	1.3233	1.3250	49
1.3369	1.3386	1.3404	1.3421	1.3439	1.3456	48
1.3581	1.3600	1.3618	1.3636	1.3655	1.3673	47
1.3805	1.3824	1.3843	1.3863	1.3882	1.3902	46
1.4040	1.4056	1.4081	1.4101	1.4122	1.4142	45
25	20	15	10	5	0	Deg.
Minutes.						

NATURAL COSECANT.

NATURAL SECANT.

Deg.	Minutes.						
	0	5	10	15	20	25	30
45	1.4142	1.4163	1.4183	1.4204	1.4225	1.4246	1.4267
46	1.4395	1.4417	1.4439	1.4461	1.4483	1.4505	1.4527
47	1.4663	1.4686	1.4709	1.4732	1.4755	1.4778	1.4802
48	1 4945	1.4969	1.4993	1.5018	1.5042	1.5067	1.5092
49	1.5242	1.5268	1.5294	1.5319	1.5345	1.5371	1.5398
50	1.5557	1.5584	1.5611	1.5639	1.5666	1.5694	1.5721
51	1.5890	1.5919	1.5947	1.5976	1.6005	1.6034	1.6064
52	1.6243	1.6273	1.6303	1.6334	1.6365	1.6396	1.6427
53	1.6616	1.6648	1.6681	1.6713	1.6746	1.6779	1.6812
54	1.7013	1.7047	1.7081	1.7116	1.7151	1.7185	1.7220
55	1.7434	1.7471	1.7507	1.7544	1.7581	1.7618	1.7655
56	1.7883	1.7921	1.7960	1.7999	1.8039	1.8078	1.8118
57	1.8361	1.8402	1.8443	1.8485	1.8527	1.8569	1.8611
58	1.8871	1.8915	1.8959	1.9004	1.9048	1.9093	1.9139
59	1.9416	1.9463	1.9510	1.9558	1.9606	1.9654	1.9703
60	2.0000	2.0050	2.0102	2.0152	2.0204	2.0256	2.0308
61	2.0627	2.0681	2.0735	2.0790	2.0846	2.0901	2.0957
62	2.1300	2.1359	2.1418	2.1477	2.1536	2.1596	2.1657
63	2.2027	2.2090	2.2153	2.2217	2.2282	2.2346	2.2411
64	2.2812	2.2880	2.2949	2.3018	2.3087	2.3158	2.3228
65	2.3662	2.3736	2.3811	2.3886	2.3961	2.4037	2.4114
66	2.4586	2.4666	2.4748	2.4829	2.4912	2.4995	2.5078
67	2.5593	2.5681	2.5770	2.5859	2.5949	2.6040	2.6131
68	2.6695	2.6791	2.6888	2.6986	2.7085	2.7185	2.7285
69	2.7904	2.8010	2.8117	2.8225	2.8334	2.8444	2.8554
70	2.9238	2.9355	2.9474	2.9593	2.9713	2.9835	2.9957
71	3.0715	3.0846	3.0977	3.1110	3.1244	3.1379	3.1515
72	3.2361	3.2506	3.2653	3.2801	3.2951	3.3102	3.3255
73	3.4203	3.4366	3.4532	3.4697	3.4867	3.5037	3.5209
74	3.6276	3.6464	3.6651	3.6840	3.7031	3.7224	3.7420
75	3.8637	3.8848	3.9061	3.9277	3.9495	3.9716	3.9939
76	4.1336	4.1578	4.1824	4.2072	4.2324	4.2579	4.2836
77	4.4454	4.4736	4.5021	4.5331	4.5604	4.5901	4.6202
78	4.8097	4.8429	4.8765	4.9106	4.9452	4.9802	5.0158
79	5.2408	5.2803	5.3205	5.3612	5.4026	5.4447	5.4874
80	5.7588	5.8067	5.8554	5.9049	5.9554	5.9963	6.0588
81	6.3924	6.4517	6.5121	6.5736	6.6363	6.7003	6.7655
82	7.1853	7.2604	7.3372	7.4156	7.4957	7.5776	7.6613
83	8.2055	8.3039	8.4046	8.5079	8.6138	8.7223	8.8337
84	9.5668	9.7010	9.8391	9.9812	10.1270	10.2780	10.4330
85	11.4740	11.6680	11.8680	12.0760	12.2010	12.5140	12.7450
86	14.3350	14.6400	14.9580	15.2900	15.6370	16.0000	16.3800
87	19.1070	19.6530	20.2300	20.8430	21.4940	22.1860	22.9250
88	28.6540	29.8990	31.2570	32.7450	34.3820	36.1910	38.2010
89	57.2990	62.5070	68.7570	76.3960	85.9460	98.2230	114.5900
Deg.	60	55	50	45	40	35	30
	Minutes.						

NATURAL COSECANT.

NATURAL SECANT.

Minutes.						Deg.
35	40	45	50	55	60	
1.4288	1.4310	1.4331	1.4352	1.4374	1.4395	44
1.4550	1.4572	1.4595	1.4617	1.4640	1.4663	43
1.4825	1.4849	1.4873	1.4897	1.4921	1.4945	42
1.5116	1.5141	1.5166	1.5192	1.5217	1.5242	41
1.5424	1.5450	1.5477	1.5503	1.5530	1.5557	40
1.5749	1.5777	1.5805	1.5833	1.5862	1.5890	39
1.6093	1.6123	1.6153	1.6182	1.6212	1.6243	38
1.6458	1.6489	1.6521	1.6552	1.6584	1.6616	37
1.6845	1.6878	1.6912	1.6945	1.6979	1.7013	36
1.7256	1.7291	1.7327	1.7362	1.7398	1.7434	35
1.7693	1.7730	1.7768	1.7806	1.7844	1.7883	34
1.8158	1.8198	1.8238	1.8279	1.8320	1.8361	33
1.8654	1.8697	1.8740	1.8783	1.8827	1.8871	32
1.9184	1.9230	1.9276	1.9322	1.9369	1.9416	31
1.9752	1.9801	1.9850	1.9900	1.9950	2.0000	30
2.0360	2.0413	2.0466	2.0519	2.0573	2.0627	29
2.1014	2.1070	2.1127	2.1185	2.1242	2.1300	28
2.1717	2.1778	2.1840	2.1902	2.1964	2.2027	27
2.2477	2.2543	2.2610	2.2676	2.2744	2.2812	26
2.3299	2.3371	2.3443	2.3515	2.3588	2.3662	25
2.4101	2.4269	2.4347	2.4426	2.4506	2.4586	24
2.5163	2.5247	2.5333	2.5419	2.5506	2.5593	23
2.6223	2.6316	2.6410	2.6504	2.6599	2.6695	22
2.7386	2.7488	2.7591	2.7694	2.7799	2.7904	21
2.8666	2.8778	2.8892	2.9006	2.9122	2.9338	20
3.0081	3.0206	3.0331	3.0458	3.0586	3.0715	19
3.1653	3.1792	3.1932	3.2074	3.2216	3.2361	18
3.3409	3.3565	3.3722	3.3881	3.4041	3.4203	17
3.5383	3.5559	3.5736	3.5915	3.6096	3.6279	16
3.7617	3.7816	3.8018	3.8222	3.8428	3.8637	15
4.0165	4.0394	4.0625	4.0859	4.1096	4.1336	14
4.3098	4.3362	4.3630	4.3901	4.4176	4.4454	13
4.6507	4.6817	4.7130	4.7448	4.7770	4.8097	12
5.0520	5.0886	5.1258	5.1636	5.2019	5.2408	11
5.5308	5.5749	5.6197	5.6653	5.7117	5.7588	10
6.1120	6.1661	6.2211	6.2772	6.3343	6.3924	9
6.8320	6.8998	6.9690	7.0396	7.1117	7.1853	8
7.7469	7.8344	7.9240	7.9971	8.1094	8.2055	7
8.9479	9.0651	9.1855	9.3092	9.4362	9.5668	6
10.5930	10.7580	10.9290	11.1040	11.2080	11.4740	5
12.9850	13.2350	13.4940	13.7630	14.0430	14.3350	4
16.7790	17.1980	17.6390	18.1030	18.5910	19.1070	3
23.7160	24.5620	25.4710	26.1500	27.5080	28.6540	2
39.9780	42.9760	45.8400	49.1140	52.8910	57.2990	1
137.5100	171.8900	229.1800	343.7700	687.5500	∞	0
25	20	15	10	5	0	Deg.
Minutes.						

NATURAL COSECANT.

CIRCUMFERENCE, AREA, AND CUBIC CONTENTS OF CIRCLES.

Diameter in inches.	Circumference in inches.	Area in square inches.	Contents of one foot in length in cubic inches.	Diameter in inches.	Circumference in inches.	Area in square inches.	Contents of one foot in length in cubic inches.	Diameter in inches.	Circumference in inches.	Area in square inches.	Contents of one foot in length in cubic inches.
1/16	0.1963	0.00307	0.03682	1 3/8	4.4320	1.48489	17.8187	2 11/16	8.4430	5.67266	68.0719
1/8	0.3927	0.01227	0.14726	1 7/16	4.5160	1.52295	19.4754	2 3/4	8.6394	5.93957	71.2749
3/16	0.5890	0.02761	0.33134	1 1/2	4.7124	1.76715	21.2057	2 13/16	8.8357	6.21262	74.5515
1/4	0.7854	0.04909	0.58905	1 9/16	4.9087	1.91748	23.0097	2 7/8	9.0321	6.49181	77.9017
5/16	0.9817	0.07670	0.92039	1 5/8	5.1051	2.07394	24.8873	2 15/16	9.2284	6.77713	81.3255
3/8	1.1781	0.11045	1.32536	1 11/16	5.3014	2.23654	26.8385	3	9.4248	7.06858	84.8230
7/16	1.3744	0.15033	1.80396	1 3/4	5.4978	2.40528	28.8634	3 1/16	9.6211	7.36618	88.3941
1/2	1.5708	0.19635	2.35619	1 13/16	5.6941	2.58015	30.9619	3 1/8	9.8175	7.66990	92.0338
9/16	1.7671	0.24850	2.98206	1 7/8	5.8905	2.76116	33.1340	3 3/16	10.0138	7.97977	95.7572
5/8	1.9635	0.30680	3.68155	1 15/16	6.0868	2.94831	35.3797	3 1/4	10.2102	8.29577	99.5492
11/16	2.1598	0.37122	4.45468	2	6.2832	3.14159	37.6991	3 5/16	10.4065	8.61790	103.4148
3/4	2.3562	0.44179	5.30143	2 1/16	6.4795	3.34101	40.0921	3 3/8	10.6029	8.94618	107.3541
13/16	2.5525	0.51849	6.22182	2 1/8	6.6759	3.54656	42.5588	3 7/16	10.7992	9.28058	111.3670
7/8	2.7489	0.60132	7.21584	2 3/16	6.8722	3.75825	45.0990	3 1/2	10.9956	9.62113	115.4535
15/16	2.9452	0.69029	8.28349	2 1/4	7.0686	3.97608	47.7129	3 9/16	11.1919	9.96781	119.6137
1	3.1416	0.78540	9.42477	2 5/16	7.2649	4.20004	50.4005	3 5/8	11.3883	10.32062	123.8475
1 1/16	3.3379	0.88664	10.63970	2 3/8	7.4613	4.43014	53.1616	3 11/16	11.5846	10.67957	128.1549
1 1/8	3.5343	0.99402	11.92820	2 7/16	7.6576	4.66637	55.9964	3 3/4	11.7810	11.04466	132.5359
1 3/16	3.7306	1.10753	13.29040	2 1/2	7.8540	4.90874	58.9049	3 13/16	11.9773	11.41588	136.9906
1 1/4	3.9270	1.22718	14.72620	2 9/16	8.0503	5.15724	61.8869	3 7/8	12.1737	11.79324	141.5189
1 5/16	4.1233	1.35297	16.23560	2 5/8	8.2467	5.41188	64.9426	3 15/16	12.3700	12.17674	146.1209

Diameter in inches.	Circumference in inches.	Area in square inches.	Contents of one foot in length in cubic inches.	Diameter in inches.	Circumference in inches.	Area in square inches.	Contents of one foot in length in cubic inches.	Diameter in inches.	Circumference in inches.	Area in square inches.	Contents of one foot in length in cubic inches.
4	12.5664	12.56637	150.7964	$5\frac{3}{8}$	16.8861	22.39064	272.2877	$7\frac{1}{2}$	23.5619	44.17865	530.1438
$4\frac{1}{16}$	12.7627	12.96214	155.5457	$5\frac{7}{16}$	17.0824	23.22140	278.6568	$7\frac{5}{8}$	23.9546	45.66354	547.9625
$4\frac{1}{8}$	12.9591	13.36404	160.3685	$5\frac{1}{2}$	17.2788	23.75829	285.0995	$7\frac{3}{4}$	24.3473	47.17298	566.0757
$4\frac{3}{16}$	13.1554	13.77208	165.2650	$5\frac{9}{16}$	17.4751	24.30132	291.6159	$7\frac{7}{8}$	24.7400	48.70696	584.4835
$4\frac{1}{4}$	13.3518	14.18625	170.2351	$5\frac{5}{8}$	17.6715	24.85049	298.2059	8	25.1327	50.26548	603.1858
$4\frac{5}{16}$	13.5481	14.60656	175.2788	$5\frac{11}{16}$	17.8678	25.40579	304.8695	$8\frac{1}{8}$	25.5254	51.84855	622.1826
$4\frac{3}{8}$	13.7445	15.03301	180.3961	$5\frac{3}{4}$	18.0642	25.96723	311.6068	$8\frac{1}{4}$	25.9181	53.45616	641.4739
$4\frac{7}{16}$	13.9408	15.46559	185.5871	$5\frac{13}{16}$	18.2605	26.53480	318.4176	$8\frac{3}{8}$	26.3108	55.08832	661.0598
$4\frac{1}{2}$	14.1372	15.90431	190.8518	$5\frac{7}{8}$	18.4569	27.10851	325.3021	$8\frac{1}{2}$	26.7035	56.74502	680.9402
$4\frac{9}{16}$	14.3335	16.34917	196.1901	$5\frac{15}{16}$	18.6532	27.68835	332.2602	$8\frac{5}{8}$	27.0962	58.42626	701.1151
$4\frac{5}{8}$	14.5299	16.80016	201.6019	6	18.8496	28.27433	339.2920	$8\frac{3}{4}$	27.4889	60.13205	721.5846
$4\frac{11}{16}$	14.7262	17.25728	207.0874	$6\frac{1}{8}$	19.2423	29.46470	353.5764	$8\frac{7}{8}$	27.8816	61.86238	742.3485
$4\frac{3}{4}$	14.9226	17.72055	212.6465	$6\frac{1}{4}$	19.6350	30.67962	368.1554	9	28.2743	63.61725	763.4070
$4\frac{13}{16}$	15.1189	18.18994	218.2793	$6\frac{3}{8}$	20.0277	31.91907	383.0289	$9\frac{1}{8}$	28.6670	65.39667	784.7600
$4\frac{7}{8}$	15.3153	18.66548	223.9853	$6\frac{1}{2}$	20.4204	33.18307	398.1969	$9\frac{1}{4}$	29.0597	67.20063	806.4076
$4\frac{15}{16}$	15.5116	19.14715	229.7658	$6\frac{5}{8}$	20.8131	34.47162	413.6594	$9\frac{3}{8}$	29.4524	69.02914	828.3496
5	15.7080	19.63495	235.6195	$6\frac{3}{4}$	21.2058	35.78470	429.4164	$9\frac{1}{2}$	29.8451	70.88248	850.5862
$5\frac{1}{16}$	15.9043	20.12890	241.5468	$6\frac{7}{8}$	21.5984	37.12233	445.4680	$9\frac{5}{8}$	30.2378	72.75978	873.1173
$5\frac{1}{8}$	16.1007	20.62897	247.5477	7	21.9911	38.48451	461.8141	$9\frac{3}{4}$	30.6305	74.66191	895.9430
$5\frac{3}{16}$	16.2970	21.13519	253.6223	$7\frac{1}{8}$	22.3838	39.87123	478.4547	$9\frac{7}{8}$	31.0232	76.58859	919.0631
$5\frac{1}{4}$	16.4934	21.64754	259.7705	$7\frac{1}{4}$	22.7765	41.28249	495.3899	10	31.4159	78.53982	942.4779
$5\frac{5}{16}$	16.6897	22.16602	265.9923	$7\frac{3}{8}$	23.1692	42.71830	512.6196	$10\frac{1}{8}$	31.8086	80.51558	966.1870

Diameter in inches.	Circum-ference in inches.	Area in square inches.	Contents of one foot in length in cubic inches.	Diameter in inches.	Circum-ference in inches.	Area in square inches.	Contents of one foot in length in cubic inches.	Diameter in inches.	Circum-ference in inches.	Area in square inches.	Contents of one foot in length in cubic inches.
10¼	32.2013	82.51589	990.1907	13	40.8407	132.73229	1592.7875	15¾	49.4801	194.82783	2337.9340
10⅜	32.5940	84.54075	1014.4890	13⅛	41.2334	135.29711	1623.5653	15⅞	49.8728	197.93261	2375.1913
10½	32.9867	86.59015	1039.0818	13¼	41.6261	137.88647	1654.5176	16	50.2655	201.06193	2416.3432
10⅝	33.3794	88.66409	1063.9691	13⅜	42.0188	140.50037	1686.0044	16⅛	50.6582	204.21579	2450.5895
10¾	33.7721	90.76257	1089.1509	13½	42.4115	143.13882	1717.5458	16¼	51.0509	207.39420	2488.7304
10⅞	34.1648	92.88560	1114.6272	13⅝	42.8042	145.80181	1749.6217	16⅜	51.4436	210.59715	2527.1658
11	34.5575	95.03318	1140.3981	13¾	43.1969	148.48934	1781.8721	16½	51.8363	213.82465	2565.8958
11⅛	34.9502	97.20529	1166.4635	13⅞	43.5896	151.20142	1814.4170	16⅝	52 2290	217.07669	2604.9203
11¼	35.3429	99.40195	1192.8235	14	43.9823	153.93804	1847.2565	16¾	52.6217	220.35327	2644.2393
11⅜	35.7356	101.62316	1219.4779	14⅛	44.3750	156.69921	1880.3905	16⅞	53.0144	223.65440	2683.8528
11½	36.1283	103.86891	1246.4269	14¼	44.7677	159.48491	1913.8190	17	53.4071	226.98007	2723.7608
11⅝	36.5210	106.13920	1273.6763	14⅜	45.1604	162.29517	1947.5420	17⅛	53.7998	230.33028	2763.9634
11¾	36.9137	108.43403	1301.2084	14½	45.5531	165.12996	1981.5596	17¼	54.1925	233.70504	2804.4605
11⅞	37.3064	110.75341	1329.0410	14⅝	45.9458	167.9[illegible]930	2015.8716	17⅜	54.5852	237.10434	2845.2521
12	37.6991	113.09734	1357.1680	14¾	46.3385	170.87319	2050.4783	17½	54.9779	240.52819	2886.3382
12⅛	38.0918	115.46580	1385.5896	14⅞	46.7312	173.78162	2085.3794	17⅝	55.3706	243.97658	2927.7189
12¼	38.4845	117.85881	1414.3057	15	47.1239	176.71459	2120.5750	17¾	55.7633	247.44951	2969.3941
12⅜	38.8772	120.27637	1443.1964	15⅛	47.5166	179.67210	2156.0652	17⅞	56.1560	250.94698	3011.3638
12½	39.2699	122.71846	1472.6216	15¼	47.9093	182.65416	2191.8499	18	56.4867	254.46901	3053.8281
12⅝	39.6626	125.18510	1502.2213	15⅜	48.3020	185.66076	2227.9291	18⅛	56.9414	258.01557	3096.1868
12¾	40.0480	127.67629	1531.9955	15½	48.6947	188.69191	2264.3029	18¼	57.3341	261.58668	3139.0401
12⅞	40.4480	130.19202	1562.3042	15⅝	49.0874	191.74760	2300.9712	18⅜	57.7268	265.18233	3182.1879

Diameter in inches.	Circumference in inches.	Area in square inches.	Contents of one foot in length in cubic inches.	Diameter in inches.	Circumference in inches.	Area in square inches.	Contents of one foot in length in cubic inches.	Diameter in inches.	Circumference in inches.	Area in square inches.	Contents of one foot in length in cubic inches.
18½	58.1195	268.80252	3225.6303	21¼	66.7588	354.65636	4255.8763	24	75.3982	452.38934	5428.6721
18⅝	58.5122	272.44726	3269.3671	21⅜	67.1515	358.84106	4306.0927	24¼	76.1836	461.86320	5542.3585
18¾	58.9049	276.11654	3313.3985	21½	67.5442	363.05030	4356.6036	24½	76.9690	471.43520	5657.2230
18⅞	59.2976	279.81037	3357.7244	21⅝	67.9369	367.28409	4407.4091	24¾	77.7544	481.10550	5773.2655
19	59.6903	283.52874	3402.3449	21¾	68.3296	371.54242	4458.5090	25	78.5398	490.87390	5890.4862
19⅛	60.0830	287.27165	3447.2598	21⅞	68.7223	375.82529	4509.9035	25¼	79.3252	500.74840	6008.8850
19¼	60.4757	291.03911	3492.4693	22	69.1150	380.13271	4561.5926	25½	80.1106	510.70520	6128.4619
19⅜	60.8684	294.83111	3537.9733	22⅛	69.5077	384.46467	4613.5761	25¾	80.8960	520.76810	6249.2168
19½	61.2611	298.64765	3583.7718	22¼	69.9004	388.82118	4665.8542	26	81.6814	530.92920	6371.1499
19⅝	61.6538	302.48874	3629.8649	22⅜	70.2931	393.20223	4718.4268	26¼	82.4668	541.18840	6494.2611
19¾	62.0465	306.35437	3676.2525	22½	70.6858	397.60782	4771.2939	26½	83.2522	551.54590	6618.5503
19⅞	62.4392	310.24455	3722.9346	22⅝	71.0785	402.03796	4824.4555	26¾	84.0376	562.00150	6744.0177
20	62.8319	314.15927	3769.9112	22¾	71.4712	406.49264	4877.9117	27	84.8230	572.55530	6870.6631
20⅛	63.2246	318.09853	3817.1823	22⅞	71.8639	410.97186	4931.6624	27¼	85.6084	583.20720	6998.4867
20¼	63.6173	322.06233	3864.7480	23	72.2566	415.47563	4985.7076	27½	86.3938	593.95740	7127.4883
20⅜	64.0100	326.05068	3912.6082	23⅛	72.6493	420.00394	5040.0473	27¾	87.1792	604.80570	7257.6681
20½	64.4026	330.06358	3960.7629	23¼	73.0420	424.55680	5094.6816	28	87.9646	615.75220	7389.0259
20⅝	64.7953	334.10102	4009.2122	23⅜	73.4347	429.13420	5149.6104	28¼	88.7500	626.79680	7521.5618
20¾	65.1880	338.16300	4057.9560	23½	73.8274	433.73614	5204.8337	28½	89.5354	637.93970	7655.2759
20⅞	65.5807	342.24952	4106.9943	23⅝	74.2201	438.36262	5260.3515	28¾	90.3208	649.18070	7790.1680
21	65.9734	346.36059	4156.3271	23¾	74.6128	443.01365	5316.1638	29	91.1062	660.51990	7926.2383
21⅛	66.3661	350.49620	4205.9544	23⅞	75.0055	447.38923	5372.2707	29¼	91.8916	671.95720	8063.4866

Diameter in inches.	Circum-ference in inches.	Area in square inches.	Contents of one foot in length in cubic inches.	Diameter in inches.	Circum-ference in inches.	Area in square inches.	Contents of one foot in length in cubic inches.	Diameter in inches.	Circum-ference in inches.	Area in square inches.	Contents of one foot in length in cubic inches.
29½	92.6770	683.4928	8201.9130	35	109.9557	962.1127	11545.3530	40½	127.2345	1288.2493	15458.9920
29¾	93.4624	695.1265	8341.5175	35¼	110.7411	975.9063	11710.8756	40¾	128.0199	1304.2027	15650.4328
30	94.2478	706.8583	8482.3002	35½	111.6265	989.7980	11877.5764	41	128.8055	1320.2543	15843.0517
30¼	95.0332	718.6884	8624.2609	35¾	112.3119	1003.7879	12045.4553	41¼	129.5907	1336.4041	16036.8487
30½	95.8186	730.6166	8767.3997	36	113.0973	1017.8760	12214.5122	41½	130.3761	1352.6520	16231.8238
30¾	96.6040	742.6431	8911.7166	36¼	113.8827	1032.0623	12384.7473	41¾	131.1615	1368.9981	16427.9770
31	97.3894	754.7676	9057.2116	36½	114.6681	1046.3467	12556.1604	42	131.9469	1385.4424	16625.3083
31¼	98.1748	766.9904	9203.8847	36¾	115.4535	1060.7293	12728.7518	42¼	132.7323	1401.9848	16823.8177
31½	98.9602	779.3113	9351.7359	37	116.2389	1075.2101	12902.5210	42½	133.5177	1418.6254	17023.5051
31¾	99.7456	791.7304	9500.7652	37¼	117.0243	1089.7890	13077.4684	42¾	134.3031	1435.3642	17224.3707
32	100.5310	804.2477	9650.9726	37½	117.8097	1104.4662	13253.5940	43	135.0885	1452.2012	17426.4144
32¼	101.3164	816.8632	9802.3581	37¾	118.5951	1119.2415	13430.8976	43¼	135.8739	1469.1363	17629.6362
32½	102.1018	829.5768	9954.9217	38	119.3805	1134.1149	13609.3793	43½	136.6593	1486.1697	17934.0361
32¾	102.8872	842.3886	10108.6634	38¼	120.1659	1149.0864	13789.0392	43¾	137.4447	1503.3012	18039.6140
33	103.6726	855.2986	10263.5832	38½	120.9513	1164.1564	13969.8771	44	138.2301	1520.5308	18246.3701
33¼	104.4580	868.3068	10419.6811	38¾	121.7367	1179.3244	14151.8931	44¼	139.0155	1537.8587	18454.3042
33½	105.2434	881.4131	10576.9571	39	122.5221	1194.5906	14335.0872	44½	139.8009	1555.2847	18663.4165
33¾	106.0288	894.6176	10735.4111	39¼	123.3075	1209.9550	14519.4595	44¾	140.5863	1572.8089	18873.7069
34	106.8142	907.9203	10895.0433	39½	124.0929	1225.4175	14705.0098	45	141.3717	1590.4313	19085.1753
34¼	107.5994	921.3211	11055.8536	39¾	124.8783	1240.9782	14891.7382	45¼	142.1571	1608.1518	19297.8219
34½	108.3849	934.8202	11217.8420	40	125.6637	1256.6371	15079.6447	45½	142.9425	1625.9705	19511.6465
34¾	109.1703	948.4174	11381.0084	40¼	126.4491	1272.3941	15268.7293	45¾	143.7279	1643.8874	19726.6493

Diameter in inches.	Circumference in inches.	Area in square inches.	Contents of one foot in length in cubic inches.	Diameter in inches.	Circumference in inches.	Area in square inches.	Contents of one foot in length in cubic inches.	Diameter in inches.	Circumference in inches.	Area in square inches.	Contents of one foot in length in cubic inches.
46	144.5133	1661.9025	19942.8301	50¾	159.4358	2022.8421	24274.1046	55½	174.3684	2419.2227	29030.6722
46¼	145.2987	1680.0158	20160.1891	51	160.2212	2042.8206	24513.8474	55¾	175.1438	2441.0666	29292.7989
46½	146.0841	1698.2272	20378.7261	51¼	161.0066	2062.8974	24755.8485	56	175.9292	2463.0086	29556.1036
46¾	146.8695	1716.5316	20598.4412	51½	161.7920	2083.0723	24996.8673	56¼	176.7146	2485.0489	29820.5865
47	147.6549	1734.9445	20819.3345	51¾	162.5774	2103.3454	25240.1443	56½	177.5100	2507.1873	30086.2474
47¼	148.4403	1753.4505	21041.4058	52	163.3628	2123.7166	25484.5995	56¾	178.2954	2529.4239	30353.0865
47½	149.2257	1772.0546	21264.6552	52¼	164.1482	2144.1861	25730.2328	57	179.0808	2551.7586	30621.1036
47¾	150.0110	1790.7569	21489.0827	52½	164.9336	2164.7537	25977.0442	57¼	179.8662	2574.1916	30890.2988
48	150.7964	1809.5574	21714.6884	52¾	165.7190	2185.4195	26225.0336	57½	180.6516	2596.7227	31160.6721
48¼	151.5818	1828.4560	21941.4721	53	166.5044	2206.1834	26474.2012	57¾	181.4370	2619.3520	31432.2235
48½	152.3672	1847.4528	22169.4339	53¼	167.2898	2227.0456	26724.5469	58	182.2224	2642.0794	31704.9531
48¾	153.1526	1866.5478	22398.5738	53½	168.0752	2248.0059	26976.6706	58¼	183.0078	2664.9051	31978.8607
49	153.9380	1885.7410	22628.8918	53¾	168.8606	2269.0644	27228.7725	58½	183.7932	2687.8289	32253.9464
49¼	154.7234	1905.0323	22860.3879	54	169.6460	2290.2210	27482.6524	58¾	184.5786	2710.8508	32530.2102
49½	155.5088	1924.4218	23093.0621	54¼	170.4314	2311.4759	27737.7105	59	185.3640	2733.9710	32807.6521
49¾	156.2942	1943.9095	23326.9144	54½	171.2168	2332.8289	27993.9467	59¼	186.1494	2757.1893	33086.2721
50	157.0796	1963.4954	23561.9448	54¾	172.0022	2354.2801	28251.3609	59½	186.9348	2780.5058	33366.0702
50¼	157.8650	1983.1794	23798.1533	55	172.7876	2375.8294	28509.9532	59¾	187.7202	2803.9205	33647.0464
50½	158.6504	2002.9617	24035.5399	55¼	173.5730	2397.4770	28769.7237	60	188.5056	2827.4334	33929.2000

SPECIFIC GRAVITIES OF MATERIALS.

GASES at 32° Fahr., and under the pressure of one atmosphere of 2116.4 lbs. on the square foot:	Weight of a cubic foot in lbs. avoirdupois.
Air	0.080728
Carbonic acid	0.12344
Hydrogen	0.005592
Oxygen	0.089256
Nitrogen	0.078596
Steam (ideal)	0.05022
Æther vapor (ideal)	0.2093
Bisulphuret-of-carbon vapour (ideal)	0 2137
Olefiant gas	0.0795

LIQUIDS at 32° Fahr. (except water, which is taken at 39°.4 Fahr.):	Weight of a cubic foot in lbs. avoirdupois.	Specific gravity, pure water = 1.
Water, pure, at 39°.4	62.425	1.000
" sea, ordinary	64.05	1.026
Alcohol, pure	49.38	0.791
" proof spirit	57.18	0.916
Æther	44.70	0.716
Mercury	848.75	13.596
Naphtha	52.94	0.848
Oil, linseed	58.68	0.940
" olive	57.12	0.915
" whale	57.62	0.923
" of turpentine	54.31	0.870
Petroleum	54.81	0.878
SOLID MINERAL SUBSTANCES, non-metallic:		
Basalt	187.3	3.00
Brick	125 to 135	2 to 2.167
Brickwork	112	1.8
Chalk	117 to 174	1.87 to 2.78
Clay	120	1.92
Coal, anthracite	100	1.602
" bituminous	77.4 to 89.9	1.24 to 1.44
Coke	62.43 to 103.6	1.00 to 1.66
Felspar	162.3	2.6
Flint	164.2	2.63

	Weight of a cubic foot in lbs. avoirdupois.	Specific gravity, pure water = 1.
SOLID MINERAL SUBSTANCES—continued:		
Glass, crown, average	156	2.5
" flint	187	3.0
" green	169	2.7
" plate	169	2.7
Granite	164 to 172	2.63 to 2.76
Gypsum	143.6	2.3
Limestone, (including marble)	169 to 175	2.7 to 2.8
" magnesian	178	2.86
Marl	100 to 119	1.6 to 1.9
Masonry	116 to 144	1.85 to 2.3
Mortar	109	1.75
Mud	102	1.63
Quartz	165	2.65
Sand (damp)	118	1.9
" (dry)	88.6	1.42
Sandstone, average	144	2.3
" various kinds	130 to 157	2.08 to 2.52
Shale	162	2.6
Slate	175 to 181	2.8 to 2.9
Trap	170	2.72
METALS, solid:		
Brass, cast	487 to 524.4	7.8 to 8.4
" wire	533	8.54
Bronze	524	8.4
Copper, cast	537	8.6
" sheet	549	8.8
" hammered	556	8.9
Gold	1186 to 1224	19 to 19.6
Iron, cast, various	434 to 456	6.95 to 7.3
" average	444	7.11
Iron, wrought, various	474 to 487	7.6 to 7.8
" average	480	7.69
Lead	712	11.4
Platinum	1311 to 1373	21 to 22
Silver	655	10.5
Steel	487 to 493	7.8 to 7.9
Tin	456 to 468	7.3 to 7.5
Zinc	424 to 449	6.8 to 7.2
TIMBER: *		
Ash	47	0.753
Bamboo	25	0.4
Beech	43	0.69

	Weight of a cubic foot in lbs. avoirdupois.	Specific gravity, pure water = 1.
TIMBER:*—continued.		
Birch	44.4	0.711
Blue-gum	52.5	0.843
Box	60	0.96
Bullet-tree	65.3	1.046
Cabacalli	56.2	0.9
Cedar of Lebanon	30.4	0.486
Chestnut	33.4	0.535
Cowrie	36.2	0.579
Ebony, West Indian	74.5	1.193
Elm	34	0.544
Fir, red pine	30 to 44	0.48 to 0.7
" spruce	30 to 44	0.48 to 0.7
" American yellow pine	29	0.46
" larch	31 to 35	0.5 to 0.56
Greenhart	62.5	1.001
Hawthorn	57	0.91
Hazel	54	0.86
Holly	47	0.76
Hornbeam	47	0.76
Laburnum	57	0.92
Lancewood	42 to 63	0.675 to 1.01
Larch. (See "fir".)		
Lignum-vitæ	41 to 83	0.65 to 1.33
Locust	44	0.71
Mahogany, Honduras	35	0.56
" Spanish	53	0.85
Maple	49	0.79
Mora	57	0.92
Oak, European	43 to 62	0.69 to 0.99
" American red	54	0.87
Poon	36	0.58
Saul	60	0.96
Sycamore	37	0.59
Teak, Indian	41 to 55	0.66 to 0.88
" African	61	0.98
Tonka	62 to 66	0.99 to 1.06
Water-gum	62.5	1.001
Willow	25	0.4
Yew	50	0.8

*The timber in every case is supposed to be dry.

WEIGHT OF A SUPERFICIAL INCH OF WROUGHT AND CAST IRON.

(From one-sixteenth to one-inch thickness.)

Thickness in inches.	Wrought Iron. Cubic foot = 480 lbs.	Cast Iron. Cubic foot = 450 lbs.
	Weight in lbs.	Weight in lbs.
$\frac{1}{16}$	0.017356	0.0163
$\frac{1}{8}$	0.0347	0.0326
$\frac{3}{16}$	0.0520	0.0489
$\frac{1}{4}$	0.0694	0.0652
$\frac{5}{16}$	0.0867	0.0815
$\frac{3}{8}$	0.1041	0.0978
$\frac{7}{16}$	0.1214	0.1141
$\frac{1}{2}$	0.1388	0.1304
$\frac{9}{16}$	0.1562	0.1467
$\frac{5}{8}$	0.1735	0.1630
$\frac{11}{16}$	0.1909	0.1793
$\frac{3}{4}$	0.2082	0.1956
$\frac{13}{16}$	0.2256	0.2119
$\frac{7}{8}$	0.2429	0.2282
$\frac{15}{16}$	0.2603	0.2445
1	0.2777	0.2608

WEIGHT PER SQUARE FOOT IN POUNDS AVOIRDUPOIS.

Thickness in inches.	Wrought Iron.	Cast Iron.	Copper, sheet.	Lead.	Zinc.
	480 lbs. per cubic foot.	450 lbs. per cubic foot.	549 lbs. per cubic foot.	712 lbs. per cubic foot.	436 lbs. per cubic foot.
$\frac{1}{16}$	2.50	2.34	2.86	3.71	2.27
$\frac{1}{8}$	5.00	4.69	5.72	7.42	4.54
$\frac{3}{16}$	7.50	7.03	8.58	11.12	6.81
$\frac{1}{4}$	10.00	9.37	11.44	14.83	9.08
$\frac{5}{16}$	12.50	11.72	14.30	18.54	11.35
$\frac{3}{8}$	15.00	14.06	17.16	22.25	13.62
$\frac{7}{16}$	17.50	16.41	20.02	25.96	15.89
$\frac{1}{2}$	20.00	18.75	22.88	29.66	18.16
$\frac{9}{16}$	22.50	21.09	25.74	33.37	20.43
$\frac{5}{8}$	25.00	23.44	28.60	37.10	22.70
$\frac{11}{16}$	27.50	25.78	31.46	40.79	24.97
$\frac{3}{4}$	30.00	28.12	34.32	44.50	27.24
$\frac{13}{16}$	32.50	30.47	37.18	48.20	29.51
$\frac{7}{8}$	35.00	32.81	40.04	51.91	31.78
$\frac{15}{16}$	37.50	35.16	42.90	55.62	34 05
1	40.00	37.50	45.75	59.33	36.33

WEIGHT OF A LINEAL FOOT OF FLAT AND SQUARE BAR IRON IN POUNDS AVOIRDUPOIS.

(480 pounds per cubic foot.)

Breadth in inches.	Thickness in inches.	Weight in lbs.	Breadth in inches.	Thickness in inches.	Weight in lbs.	Breadth in inches.	Thickness in inches.	Weight in lbs.
¼	⅛	0.104	1½	1	5.000	2¼	¼	1.875
"	¼	0.208	"	1⅛	5.625	"	⅜	2.813
½	⅛	0.208	"	1¼	6.250	"	½	3.750
"	¼	0.416	"	1⅜	6.874	"	⅝	4.687
"	½	0.832	"	1½	7.500	"	¾	5.624
¾	⅛	0.312	1¾	⅛	0.739	"	⅞	6.562
"	¼	0.624	"	¼	1.459	"	1	7.500
"	⅜	0.937	"	⅜	2.187	"	1⅛	8.437
"	½	1.249	"	½	2.916	"	1¼	9.374
"	⅝	1.562	"	⅝	3.646	"	1⅜	10.310
"	¾	1.874	"	¾	4.375	"	1½	11.250
1	⅛	0.416	"	⅞	5.103	"	1⅝	12.190
"	¼	0.833	"	1	5.833	"	1¾	13.120
"	⅜	1.249	"	1⅛	6.562	"	1⅞	14.060
"	½	1.667	"	1¼	7.291	"	2	15.000
"	⅝	2.089	"	1⅜	8.020	"	2⅛	15.940
"	¾	2.500	"	1½	8.750	"	2¼	17.810
"	⅞	2.916	"	1⅝	9.478	2½	⅛	1.041
"	1	3.333	"	1¾	10.930	"	¼	2.089
1¼	⅛	0.521	2	⅛	0 833	"	⅜	3.125
"	¼	1.041	"	¼	1.667	"	½	4.166
"	⅜	1.562	"	⅜	2.500	"	⅝	5.208
"	½	2.089	"	½	3.333	"	¾	6.250
"	⅝	2.603	"	⅝	4.166	"	⅞	7.291
"	¾	3.124	"	¾	5.000	"	1	8.333
"	⅞	3.646	"	⅞	5.833	"	1⅛	9.398
"	1	4.166	"	1	6.666	"	1¼	10.410
"	1⅛	4.687	"	1⅛	7.500	"	1⅜	11.460
"	1¼	5.728	"	1¼	8.333	"	1½	12.500
1½	⅛	0.624	"	1⅜	9.156	"	1⅝	13.540
"	¼	1.250	"	1½	10.000	"	1¾	14.580
"	⅜	1.875	"	1⅝	10.830	"	1⅞	15.620
"	½	2.500	"	1¾	11.660	"	2	16.660
"	⅝	3.125	"	1⅞	12.500	"	2⅛	17.710
"	¾	3.750	"	2	13.330	"	2¼	18.750
"	⅞	4.375	2¼	⅛	0.937	"	2½	20.820

Breadth in inches.	Thickness in inches.	Weight in lbs.	Breadth in inches.	Thickness in inches.	Weight in lbs.	Breadth in inches.	Thickness in inches.	Weight in lbs.
2½	2⅜	19.800	3¼	2	21.660	4	2	26.660
2¾	⅛	1.146	"	2¼	24.370	"	2¼	30.000
"	¼	2.292	"	2½	27.080	"	2½	33.330
"	⅜	3.437	"	2¾	29.790	"	2¾	36.660
"	½	4.583	"	3	32.500	"	3	40.000
"	⅝	5.729	"	3¼	24.200	"	3¼	43.330
"	¾	6.874	3½	¼	2.916	"	3½	46.660
"	⅞	8.020	"	½	5.833	"	3¾	50.000
"	1	9.154	"	¾	8.750	"	4	53.330
"	1⅛	10.310	"	1	11.660	4¼	¼	3.541
"	1¼	11.460	"	1¼	14.580	"	½	7.082
"	1⅜	12.600	"	1½	17 500	"	¾	10.620
"	1½	13.750	"	1¾	20.430	"	1	14.160
"	1⅝	14.900	"	2	23.330	"	1¼	16.800
"	1¾	16.030	"	2¼	26.250	"	1½	21.330
"	1⅞	17.190	"	2½	29.160	"	1¾	24.780
"	2	18.330	"	2¾	32.080	"	2	28.330
"	2⅛	19.480	"	3	35.000	"	2¼	31.870
"	2¼	20.620	"	3¼	37.910	"	2½	35.410
"	2⅜	21.770	"	3½	40.830	"	2¾	38.950
"	2½	22.910	3¾	¼	3.125	"	3	42.500
"	2⅝	24.060	"	½	6.250	"	3¼	46.030
"	2¾	25.200	"	¾	9.375	"	3½	49.570
3	¼	2.500	"	1	12.500	"	3¾	53.120
"	½	5.000	"	1¼	15.620	"	4	56.660
"	¾	7.500	"	1½	18.750	"	4¼	60.200
"	1	10.000	"	1¾	21.870	4½	¼	3.750
"	1¼	12.500	"	2	25.000	"	½	7.500
"	1½	15.000	"	2¼	28.120	"	¾	11.250
"	1¾	17.500	"	2½	31.250	"	1	15.000
"	2	20.000	"	2¾	34.370	"	1¼	18.750
"	2¼	22.500	"	3	37.500	"	1½	22.500
"	2½	25.000	"	3¼	40.620	"	1¾	26.250
"	2¾	27.500	"	3½	43.750	"	2	30.000
"	3	30.000	"	3¾	46.860	"	2¼	33.750
3¼	¼	2.708	4	¼	3.330	"	2½	37.500
"	½	5.416	"	½	6.660	"	2¾	41.250
"	¾	8.124	"	¾	10.000	"	3	45.000
"	1	10.830	"	1	13.330	"	3¼	48.750
"	1¼	13.500	"	1¼	16.660	"	3½	52.500
"	1½	16.250	"	1½	20.000	"	3¾	56.250
"	1¾	18.950	"	1¾	23.330	"	4	60.000

Breadth in inches.	Thickness in inches.	Weight in lbs.	Breadth in inches.	Thickness in inches.	Weight in lbs.	Breadth in inches.	Thickness in inches.	Weight in lbs.
4½	4¼	63.750	5¼	½	8.753	5¾	¼	4.788
"	4½	67.500	"	¾	13.130	"	½	9.587
4¾	¼	3.953	"	1	17.500	"	¾	14.370
"	½	7.910	"	1¼	21.870	"	1	19.160
"	¾	11.860	"	1½	26.250	"	1¼	23.950
"	1	15.830	"	1¾	30.620	"	1½	28.750
"	1¼	19.760	"	2	35.000	"	1¾	33.540
"	1½	23.750	"	2¼	39.370	"	2	38.330
"	1¾	27.700	"	2½	43.750	"	2¼	43.120
"	2	31.670	"	2¾	48.110	"	2½	47.910
"	2¼	35.620	"	3	52.500	"	2¾	52.700
"	2½	39.580	"	3¼	56.680	"	3	57.500
"	2¾	43.540	"	3½	61.250	"	3¼	62.300
"	3	47.500	"	3¾	65.620	"	3½	67.080
"	3¼	51.460	"	4	70.000	"	3¾	71.860
"	3½	55.410	"	4¼	74.370	"	4	76.650
"	3¾	59.370	"	4½	78.750	"	4¼	81.450
"	4	63.330	"	4¾	83.110	"	4½	86.240
"	4¼	67.290	"	5	87.500	"	4¾	91.030
"	4½	71.250	"	5¼	91.860	"	5	95.820
"	4¾	75.200	5½	¼	4.587	"	5¼	100.600
5	¼	4.166	"	½	9.164	"	5½	105.400
"	½	8.330	"	¾	13.750	"	5¾	119.700
"	¾	12.500	"	1	18.330	6	½	10.000
"	1	16.660	"	1¼	22.900	"	1	20.000
"	1¼	20.830	"	1½	27.500	"	1½	30.000
"	1½	25.000	"	1¾	32.080	"	2	40.000
"	1¾	29.160	"	2	36.660	"	2½	50.000
"	2	33.330	"	2¼	41.250	"	3	60.000
"	2¼	37.500	"	2½	45.830	"	3½	70.000
"	2½	41.660	"	2¾	50.310	"	4	80.000
"	2¾	45.830	"	3	55.000	"	4½	90.000
"	3	50.000	"	3¼	59.570	"	5	100.000
"	3¼	54.160	"	3½	64.160	"	5½	110.000
"	3½	58.330	"	3¾	68.740	"	6	120.000
"	3¾	62.500	"	4	73.330	6½	½	10.830
"	4	66.660	"	4¼	77.910	"	1	21.660
"	4¼	70.830	"	4½	82.500	"	1½	32.500
"	4½	75.000	"	4¾	87.080	"	2	43.330
"	4¾	79.160	"	5	91.560	"	2½	54.160
"	5	83.330	"	5¼	96.240	"	3	65.000
5¼	¼	4.376	"	5½	100.600	"	3½	75.830

Breadth in inches.	Thickness in inches.	Weight in lbs.	Breadth in inches.	Thickness in inches.	Weight in lbs.	Breadth in inches.	Thickness in inches.	Weight in lbs.
6½	4	86.66	8	4	106.60	9	8½	255.00
"	4½	97.50	"	4½	120.00	"	9	270.00
"	5	108.30	"	5	133.30	9½	½	15.83
"	5½	119.10	"	5½	146.60	"	1	31.66
"	6	130.00	"	6	160.00	"	1½	47.50
"	6½	140.80	"	6½	173.30	"	2	63.33
7	½	11.66	"	7	186.60	"	2½	79.16
"	1	23.33	"	7½	200.00	"	3	95.00
"	1½	35.00	"	8	213.30	"	3½	110.80
"	2	46.66	8½	½	14.16	"	4	126.60
"	2½	58.33	"	1	28.33	"	4½	142.50
"	3	70.00	"	1½	42.48	"	5	158.30
"	3½	81 66	"	2	56.66	"	5½	174.10
"	4	93.33	"	2½	70.83	"	6	190.00
"	4½	105.00	"	3	85.00	"	6½	205.80
"	5	116.60	"	3½	99.16	"	7	221.60
"	5½	128.30	"	4	113.30	"	7½	237.60
"	6	140.00	"	4½	127.50	"	8	253.30
"	6½	151.60	"	5	141.60	"	8½	269.10
"	7	163.30	"	5½	155.80	"	9	285.00
7½	½	12.50	"	6	170.00	"	9½	300.80
"	1	25.00	"	6½	184.10	10	½	16.66
"	1½	37.50	"	7	198.30	"	1	33.33
"	2	50.00	"	7½	212.50	"	1½	50.00
"	2½	62.50	"	8	226.60	"	2	66.66
"	3	75.00	"	8½	240.70	"	2½	83.33
"	3½	87.50	9	½	15.00	"	3	100.00
"	4	100.00	"	1	30.00	"	3½	116.60
"	4½	112.50	"	1½	45.00	"	4	133.30
"	5	125.00	"	2	60.00	"	4½	150.00
"	5½	137.50	"	2½	75.00	"	5	166.60
"	6	150.00	"	3	90.00	"	5½	183.30
"	6½	162.50	"	3½	105.00	"	6	200.00
"	7	175.00	"	4	120.00	"	6½	216.60
"	7½	187.50	"	4½	135.00	"	7	233.30
8	½	13.33	"	5	150.00	"	7½	250.00
"	1	26.66	"	5½	165.00	"	8	266.60
"	1½	40.00	"	6	180.00	"	8½	283.30
"	2	53.33	"	6½	195.00	"	9	300.00
"	2½	66.66	"	7	210.00	"	9½	316.60
"	3	80.00	"	7½	225.00	"	10	333.30
"	3½	93.33	"	8	240.00	10½	½	17.50

Breadth in inches.	Thickness in inches.	Weight in lbs.	Breadth in inches.	Thickness in inches.	Weight in lbs.	Breadth in inches.	Thickness in inches.	Weight in lbs.
10½	1	35.00	11	1½	55.00	11½	1½	57.50
"	1½	52.50	"	2	73.33	"	2	76.66
"	2	70.00	"	2½	91.56	"	2½	95.83
"	2½	87.50	"	3	110.00	"	3	115.00
"	3	105.00	"	3½	128.30	"	3½	134.10
"	3½	122.50	"	4	146.60	"	4	153.30
"	4	140.00	"	4½	165.00	"	4½	172.50
"	4½	157.50	"	5	183.30	"	5	191.60
"	5	175.00	"	5½	201.60	"	5½	210.80
"	5½	192.50	"	6	220.00	"	6	230.00
"	6	210.00	"	6½	238.30	"	6½	249.10
"	6½	227.50	"	7	256.60	"	7	268.30
"	7	245.00	"	7½	275.00	"	7½	287.50
"	7½	262.50	"	8	293.30	"	8	306.60
"	8	280.00	"	8½	311.60	"	8½	325.80
"	8½	297.50	"	9	330.00	"	9	345.00
"	9	315.00	"	9½	348.30	"	9½	364.10
"	9½	332.50	"	10	366.60	"	10	383.30
"	10	350.00	"	10½	385.00	"	10½	402.50
"	10½	367.50	"	11	403.30	"	11	421.60
11	½	18.33	11½	½	19.16	"	11½	440.70
"	1	36.66	"	1	38.33	12	12	480.00

WEIGHT OF A LINEAL FOOT OF ROLLED ROUND IRON IN POUNDS AVOIRDUPOIS.

(480 pounds per cubic foot.)

Diameter in inches.	Weight in lbs.	Diameter in inches.	Weight in lbs.	Diameter in inches.	Weight in lbs.	Diameter in inches.	Weight in lbs.
1/16	0.010	2 3/8	14.77	5 5/8	82.79	8 7/8	206.2
1/8	0.041	2 1/2	16.36	5 3/4	86.52	9	212.2
3/16	0.091	2 5/8	18.04	5 7/8	90.34	9 1/8	218.0
1/4	0.163	2 3/4	19.80	6	94.26	9 1/4	223.9
5/16	0.255	2 7/8	21.64	6 1/8	98.18	9 3/8	230.1
3/8	0.368	3	23.56	6 1/4	102.20	9 1/2	236.2
7/16	0.501	3 1/8	25.56	6 3/8	106.40	9 5/8	242.5
1/2	0.655	3 1/4	27.64	6 1/2	110 60	9 3/4	248.9
9/16	0.828	3 3/8	29.82	6 5/8	114.90	9 7/8	255.2
5/8	1.022	3 1/2	32.07	6 3/4	119.30	10	261.7
11/16	1.237	3 5/8	34.39	6 7/8	123.70	10 1/8	268.4
3/4	1.473	3 3/4	36.81	7	128.30	10 1/4	275.0
13/16	1.728	3 7/8	39.30	7 1/8	132.90	10 3/8	281.8
7/8	2.004	4	41.88	7 1/4	137.60	10 1/2	288.6
15/16	2.301	4 1/8	44.57	7 3/8	142.30	10 5/8	295.6
1	2.618	4 1/4	47.28	7 1/2	147.30	10 3/4	302.5
1 1/8	3.310	4 3/8	50.10	7 5/8	152.20	10 7/8	309.5
1 1/4	4.094	4 1/2	53.02	7 3/4	157 20	11	316.8
1 3/8	4.950	4 5/8	56.03	7 7/8	162.40	11 1/8	323.9
1 1/2	5.885	4 3/4	59.05	8	167.50	11 1/4	331.3
1 5/8	6.911	4 7/8	62.17	8 1/8	172.80	11 3/8	338.7
1 3/4	8.018	5	65.49	8 1/4	178.20	11 1/2	346.2
1 7/8	9.205	5 1/8	68.71	8 3/8	183.60	11 5/8	353.7
2	10.470	5 1/4	72.13	8 1/2	189.10	11 3/4	361.5
2 1/8	11.820	5 3/8	75.65	8 5/8	194.80	11 7/8	369.1
2 1/4	13.250	5 1/2	79.17	8 3/4	200.40	12	376.9

BOLTS, NUTS, AND HEADS.

(Whitworth's Proportions.)

Weight in lbs. of Heads and Nuts.

Diameter of bolt in in.	Hexagonal.		Square.		Hexagonal.		Square.	
	Head.	Nut.	Head.	Nut.	Two Heads.	Head & Nut.	Two Heads.	Head & Nut.
$\frac{1}{4}$	0.008	0.005	0.022	0.019	0.017	0.013	0.044	0.041
$\frac{5}{16}$	0.014	0.007	0.027	0.021	0.029	0.022	0.055	0.048
$\frac{3}{8}$	0.029	0.017	0.061	0.049	0.057	0.046	0.122	0.110
$\frac{7}{16}$	0.059	0.040	0.069	0.050	0.119	0.101	0.138	0.119
$\frac{1}{2}$	0.068	0.041	0.104	0.076	0.136	0.109	0.208	0.181
$\frac{9}{16}$	0.104	0.065	0.157	0.118	0.208	0.169	0.315	0.276
$\frac{5}{8}$	0.151	0.097	0.246	0.193	0.302	0.248	0.493	0.440
$\frac{3}{4}$	0.254	0.161	0.362	0.269	0.508	0.415	0.724	0.631
$\frac{7}{8}$	0.367	0.219	0.551	0.408	0.734	0.586	1.102	0.959
1	0.546	0.326	0.683	0.463	1.092	0.872	1.366	1.146
$1\frac{1}{8}$	0.724	0.411	1.109	0.797	1.448	1.135	2.217	1.906
$1\frac{1}{4}$	1.060	0.630	1.400	0.971	2.120	1.690	2.800	2.371
$1\frac{3}{8}$	1.330	0.759	1.949	1.379	2.660	2.088	3.898	3.328
$1\frac{1}{2}$	1.840	1.098	2.625	1.883	3.680	2.938	5.250	4.508
$1\frac{5}{8}$	2.460	1.517	3.135	2.192	4.920	3.977	6.270	5.327
$1\frac{3}{4}$	2.920	1.742	3.704	2.532	5.840	4.662	7.409	6.236
$1\frac{7}{8}$	3.440	1.991	4.725	3.276	6.880	5.431	9.450	8.001
2	4.370	2.611	6.384	4.625	8.740	6.981	12.77	11.00
$2\frac{1}{4}$	6.150	3.645	8.858	6.353	12.30	9.795	17.71	15.21
$2\frac{1}{2}$	8.480	5.045	11.91	8.476	16.96	13.52	23.82	20.39
$2\frac{3}{4}$	11.32	6.747	15.59	9.019	22.64	18.06	31.18	24.61
3	14.72	8.783	21.00	15.06	29.44	23.50	42.00	36.06

WEIGHT IN POUNDS OF ROUND IRON FOR

Diameter in inches.	Length in inches.									
	⅛	¼	⅜	½	⅝	¾	⅞	1	2	3
¼	0.002	0.003	0.005	0.007	0.008	0.010	0.012	0.014	0.027	0.041
5/16	0.003	0.005	0.008	0.011	0.013	0.016	0.019	0.021	0.043	0.064
⅜	0.004	0.007	0.011	0.015	0.019	0.023	0.027	0.031	0.062	0.093
7/16	0.005	0.010	0.016	0.021	0.026	0.031	0.036	0.042	0.084	0.126
½	0.007	0.014	0.021	0.027	0.034	0.041	0.048	0.055	0.110	0.166
9/16	0.009	0.017	0.026	0.035	0.043	0.052	0.061	0.069	0.139	0.208
⅝	0.011	0.022	0.032	0.043	0.054	0.065	0.076	0.087	0.174	0.261
¾	0.015	0.031	0.046	0.062	0.077	0.093	0.108	0.124	0.249	0.373
⅞	0.021	0.042	0.063	0.084	0.105	0.126	0.148	0.170	0.338	0.508
1	0.027	0.055	0.083	0.110	0.138	0.165	0.193	0.221	0.442	0.663
1⅛	0.035	0.070	0.105	0.140	0.185	0.210	0.245	0.280	0.560	0.840
1¼	0.043	0.087	0.131	0.173	0.217	0.262	0.304	0.347	0.695	1.043
1⅜	0.053	0.104	0.157	0.209	0.261	0.314	0.366	0.418	0.836	1.255
1½	0.062	0.124	0.186	0.249	0.311	0.373	0.435	0.497	0.995	1.493
1⅝	0.072	0.143	0.215	0.287	0.358	0.430	0.502	0.584	1.168	1.752
1¾	0.084	0.168	0.253	0.337	0.421	0.506	0.590	0.677	1.354	2.032
1⅞	0.097	0.194	0.291	0.389	0.486	0.583	0.680	0.778	1.555	2.333
2	0.111	0.221	0.332	0.442	0.553	0.663	0.774	0.884	1.770	2.654
2¼	0.140	0.280	0.420	0.560	0.700	0.840	0.980	1.120	2.240	3.360
2½	0.174	0.347	0.521	0;695	0.869	1.042	1.216	1.390	2.781	4.172
2¾	0.209	0.418	0.627	0.836	1.045	1.254	1.463	1.673	3.346	5.019
3	0.250	0.500	0.750	1.000	1.250	1.500	1.750	1.990	3.981	5.972

EXAMPLE.—Required, the weight of a bolt 1¼ inches diameter, 4 inches between inside of head and nut.

Weight of bolt = 1.39
Weight of square head = 1.40
Weight of hexagonal nut = 1.06 taken as a hexagonal head

Ans. 3.85 lbs.

BOLTS, ETC., BETWEEN HEAD AND NUT.

Diameter in inches.	Length in inches.								
	4	5	6	7	8	9	10	11	12
1/4	0.055	0.069	0.082	0.096	0.110	0.124	0.137	0.151	0.165
5/16	0.086	0.107	0.128	0.150	0.171	0.192	0.214	0.235	0.257
3/8	0.124	0.155	0.186	0.217	0.248	0.279	0.311	0.342	0.373
7/16	0.167	0.209	0.251	0.293	0.335	0.377	0.419	0.461	0.503
1/2	0.221	0.276	0.331	0.386	0.442	0.497	0.552	0.607	0.663
9/16	0.277	0.347	0.416	0.486	0.555	0.624	0.694	0.763	0.833
5/8	0.347	0.434	0.521	0.608	0.695	0.782	0.869	0.956	1.043
3/4	0.497	0.622	0.746	0.871	0.995	1.119	1.244	1.368	1.493
7/8	0.677	0.846	1.016	1.185	1.354	1.524	1.693	1.862	2.032
1	0.884	1.105	1.326	1.548	1.769	1.990	2.211	2.432	2.654
1 1/8	1.120	1.400	1.680	1.960	2.240	2.520	2.800	3.080	3.360
1 1/4	1.390	1.738	2.085	2.433	2.781	3.128	3.476	3.823	4.172
1 3/8	1.673	2.091	2.510	2.928	3.346	3.765	4.182	4.601	5.019
1 1/2	1.990	2.488	2.985	3.483	3.981	4.478	4.976	4.973	5.972
1 5/8	2.336	2.920	3.504	4.088	4.673	5.257	5.841	6.425	7.010
1 3/4	2.709	3.386	4.064	4.741	5.418	6.096	6.773	7.450	8.128
1 7/8	3.111	3.888	4.666	5.334	6.221	6.999	7.777	8.547	9.333
2	3.538	4.423	5.307	6.192	7.077	7.961	8.846	9.730	10.610
2 1/4	4.480	5.600	6.720	7.840	8.960	10.080	11.200	12.320	13.440
2 1/2	5.562	6.953	8.343	9.734	11.120	12.510	13.910	15.290	16.690
2 3/4	6.692	8.365	10.040	11.710	13.380	15.060	16.730	18.400	20.070
3	7.962	9.953	11.940	13.930	15.920	17.910	19.910	21.890	23.890

WEIGHT OF MATERIALS USED IN BUILDING.

(Per square foot from one inch thickness to a cubic foot.)

Stones, Earths, &c.

Thickness in inches.	Asphaltum, average.	Basalts, average.	Brick.			Plaster of Paris.	Common gravel.	Limestone.	Marble, average.	Mortar.	Mud.	Oyster shell.
			Average.	Fire.	In cement or mortar.							
1	6.58	14.58	8.50	11.41	9.33	6.12	9.08	16.5	14.08	8.16	8.5	10.83
2	13.16	29.16	17.00	22.83	18.66	12.25	18.16	33.0	28.16	16.33	17.0	21.66
3	19.74	43.74	25.50	34.24	28.00	18.36	27.24	49.5	42.25	24.50	25.5	32.49
4	26.32	58.32	34.00	45.66	37.33	24.50	36.33	66.0	56.32	32.66	34.0	43.33
5	32.90	72.90	42.50	57.08	46.66	30.61	45.41	82.5	70.40	40.83	42.5	54.16
6	39.48	87.48	51.00	68.50	56.00	36.74	54.50	99.0	84.48	49.00	51.0	65.00
7	46.06	102.06	59.50	80.00	65.33	42.86	63.60	115.5	98.56	57.16	59.5	75.83
8	52.64	116.64	68.00	91.32	74.66	49.00	72.66	132.0	112.64	65.32	68.0	86.66
9	59.22	131.22	76.50	102.75	84.00	55.10	81.75	148.5	126.72	72.50	76.5	97.50
10	65.80	145.80	85.00	114.16	93.33	61.23	90.83	165.0	140.80	81.66	85.0	108.33
11	72.38	160.38	93.50	125.60	102.66	67.35	99.13	181.5	154.90	89.82	93.5	119.16
12	79.00	175.00	102.00	137.00	112.00	73.50	109.00	198.0	169.00	98.00	102.0	130.00

Stones, Earths, &c.

Thickness in inches.	Portland cement.	Chalk.	Clay.	Clay with gravel.	Concrete, average.	Earth, common soil.	Glass, (window.)	Common sand.	Slate.	Granite.		Rain water.
										Patapsco.	Susquehanna.	
1	6.75	11.16	10.0	12.91	10.41	11.41	13.75	8.66	12.25	13.75	14.08	5.21
2	13.50	22.33	20.0	25.82	20.83	22.83	27 50	17.33	24.50	27.50	28.16	10.42
3	20.25	33.50	30.0	38.73	31.25	34.25	41.25	26.00	36.75	41.25	42.24	15.62
4	27.00	44.66	40.0	51.64	41.66	45.66	55.00	34.66	49.00	55.00	56.32	20.83
5	33.75	55.83	50.0	64 55	52.08	57 08	68 75	43.33	61.25	68.75	70.40	26.04
6	40.50	67.00	60.0	77.46	64.50	68.50	82.50	52.00	73.50	82.50	84.48	31.24
7	47.25	78.16	70.0	90.37	73.00	80.00	96.25	60.66	85.75	96.25	98.56	36.45
8	54.00	89.33	80.0	103.28	83.32	91.32	110.00	69.22	98.00	110.00	112.64	41.66
9	60.75	100.50	90.0	116.19	93.75	102.75	123.75	80.00	110.25	123.75	126.72	46 87
10	67.50	111.66	100.0	129.10	104.16	114.16	137.50	86.66	122.50	137.50	140.80	52.08
11	74.25	122.83	110.0	142.01	114.57	125 57	150.25	95.32	134.75	150.25	154.88	57.28
12	81.00	134.00	120.0	155.00	125.00	137.00	165.00	104.00	147.00	165.00	169.00	62.50

DIVISIONS OF A FOOT EXPRESSED IN EQUIVALENT DECIMALS.

	INCHES.											
	0	1	2	3	4	5	6	7	8	9	10	11
0	.00000	.08333	.16666	.25	.33333	.41666	.5	.58333	.66666	.75	.83333	.91666
$\frac{1}{16}$	.00521	.08854	.17187	.25521	.33854	.42187	.50521	.58854	.67187	.75521	.83854	.92187
$\frac{1}{8}$	.01041	.09374	.17707	.26041	.34374	.42707	.51041	.59374	.67707	.76041	.84374	.92707
$\frac{3}{16}$	.01562	.09895	.18228	.26562	.34895	.43228	.51562	.59895	.68228	.76562	.84895	.93228
$\frac{1}{4}$	.02083	.10416	.18750	.27083	.35416	.43750	.52083	.60416	.68750	.77083	.85416	.93750
$\frac{5}{16}$	.02604	.10937	.19270	.27604	.35937	.44270	.52604	.60937	.69270	.77604	.85936	.94270
$\frac{3}{8}$	.03125	.11458	.19791	.28125	.36458	.44791	.53125	.61458	.69791	.78125	.86458	.94791
$\frac{7}{16}$	.03646	.11979	.20312	.28646	.36979	.45312	.53646	.61979	.70312	.78646	.86979	.95312
$\frac{1}{2}$	.04166	.12500	.20832	.29166	.37500	.45833	.54166	.62500	.70832	.79166	.87500	.95833
$\frac{9}{16}$	.04687	.13020	.21353	.29687	.38020	.46353	.54687	.63020	.71353	.79687	.88020	.96353
$\frac{5}{8}$	.05208	.13541	.21874	.30208	.38541	.46875	.55208	.63541	.71874	.80208	.88541	.96875
$\frac{11}{16}$	.05729	.14062	.22395	.30729	.39062	.47395	.55729	.64062	.72395	.80729	.89062	.97395
$\frac{3}{4}$	.06250	.14583	.22916	.31250	.39583	.47916	.56250	.64583	.72916	.81250	.89583	.97916
$\frac{13}{16}$	.06771	.15104	.23437	.31771	.40104	.48437	.56771	.65104	.73437	.81771	.90104	.98437
$\frac{7}{8}$	.07292	.15625	.23958	.32292	.40625	.48958	.57292	.65625	.73958	.82292	.90625	.98958
$\frac{15}{16}$	.07813	.16146	.24479	.32813	.41146	.49479	.57813	.66146	.74479	.82813	.91146	.99479

To find the divisions of an inch expressed in decimals, multiply the above equivalents by 12; for instance, $4\frac{3}{16}$ inches in decimals of a foot = .34895 × 12 = 4.1874 inches.

TABLE FOR COMPARING MEASURES AND WEIGHTS OF DIFFERENT COUNTRIES.

Weights.

United States and England.	Prussia.	Austria.	Baden and Switzerland.	France.
Pound.	Pound, *Z. V.*	Pound.	Pound.	Kilogra'e.
1	0.9072	0.8100		0.4536
1.1023	1	0.8928	Same as	0.5000
1.2346	1.1200	1	Prussia.	0.5600
1.2346	1.1200	0.9999		0.5600
2.2046	2.0000	1.7857		1

Measures of Length.

Foot.	Foot.	Foot.	Foot.	Meter.
= 12 inches.	= 12 inches.	= 12 inches.	= 10 inches.	= 100 Centi.
1	0.9711	0.9642	1..0160	0.3048
1.0297	1	0.9929	1.0462	0.3138
1.0371	1.0072	1	1.0537	0.3161
0.9843	0.9559	0.9490	1	0.3000
3.2809	3.1862	3.1635	3.3333	1

Measures of Surface—Square Measure.

Square foot.	Square foot.	Square foot.	Square foot.	Sq. Meter.
1	0.9431	0.9297	1.0322	0.0929
1.0603	1	0.9858	1.0945	0.0985
1.0756	1.0144	1	1.1103	0.0999
0.9688	0.9137	0.9007	1	0.0900
10.7643	10.1519	10.0074	11.1111	1

Cubic Measure.

United States and England.	Prussia.	Austria.	Baden and Switzerland.	France.
Cubic foot.	Cubic foot.	Cubic foot.	Cubic foot.	Cubic meter
1	0.9159	0.8964	1.0487	0.0283
1.0918	1	0.9787	1.1450	0.0309
1.1156	1.0217	1	1.1699	0.0316
0.9535	0.8733	0.8548	1	0.0270
35.3166	32.3459	31.6578	37.0370	1

Weight per Unit of Length.

Lbs. per lineal foot.	Lbs. per lineal foot.	Lbs. per lineal foot.	Lbs. per lineal foot.	Kil. per lineal meter
1	0.9342	0.8400	0.8929	1.4882
1.0705	1	0.8993	0.9559	1.5931
1.1904	1.1120	1	1.0629	1.7716
1.1199	1.0462	1.9408	1	1.6667
0.6720	0.6277	0.5645	0.6000	1

Weight per Unit of Surface.

Lbs. per square inch.	Lbs. per square inch.	Lbs. per square inch.	Lbs. per square inch.	Kil. per square cent.
1	0.9619	0.8712	1.2656	0.0703
1.0396	1	0.9057	1.3157	0.0731
1.1478	1.1041	1	1.4526	0.0807
0.7902	0.7601	0.6884	1	0.0556
14.2223	13.6811	12.3910	18.0000	1

RESISTANCE TO CROSS-BREAKING.

To Cut the Strongest and Stiffest Rectangular Beam from a Log.

Fig. 308. (Strongest.)

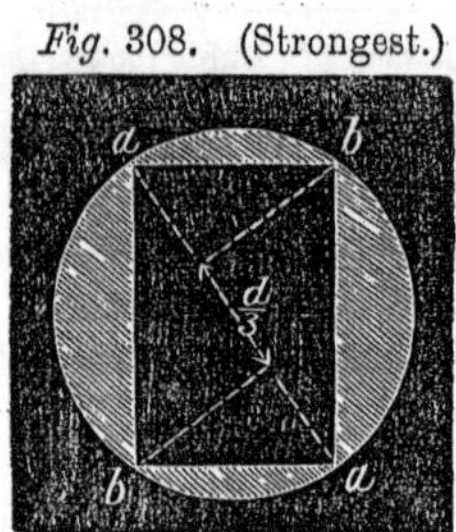

The diameter $aa = d$, divided into three equal parts, with perpendiculars $\frac{1}{3}$ d from a erected thereon, intersecting the circle at b, will give section for greatest capacity.

Fig. 309. (Stiffest.)

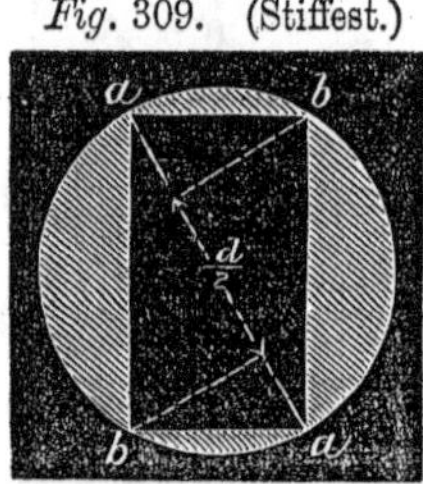

The diameter $aa = d$, divided into four equal parts, with perpendiculars $\frac{1}{4}$ d from a erected thereon, intersecting the circle at b, will give section with least deflection, but less capacity than *Fig.* 308.

INDEX.

www.ingramcontent.com/pod-product-compliance
Lightning Source LLC
LaVergne TN
LVHW010250110826
845151LV00004B/1431